# Analysis of Messy Data

## VOLUME 2: NONREPLICATED EXPERIMENTS

**George A. Milliken**
**Dallas E. Johnson**

VAN NOSTRAND REINHOLD
—————————— New York

Van Nostrand Reinhold
115 Fifth Avenue
New York, New York 10003

Van Nostrand Reinhold International Company Limited
11 New Fetter Lane
London EC4P 4EE, England

Van Nostrand Reinhold
480 La Trobe Street
Melbourne, Victoria 3000, Australia

Nelson Canada
1120 Birchmount Road
Scarborough, Ontario MIK 5G4, Canada

16 15 14 13 12 11 10 9 8 7 6 5 4 3 2

**Library of Congress Cataloging in Publication Data**

(Revised for vol. 2)

Milliken, George A., 1943–
   Analysis of messy data.

   Vol. 2 has imprint: New York : Van Nostrand Reinhold
   Includes bibliographies and indexes.
   Content: v. 1. Designed experiments—v. 2. Nonreplicated experiments.
   1. Analysis of variance.   2. Experimental design.
3. Sampling (Statistics)   I. Johnson, Dallas, E.,
1938–    . II. Title.
QA279.M48   1984         519.5'352         84-839
ISBN 0-534-02713-X (v. 1)
ISBN 0-442-24408-8 (v. 2)

# Contents

# Preface

This volume considers the analysis and design of nonreplicated experiments. It is very important that such experiments be designed efficiently and analyzed correctly, since these kinds of experiments are often much more expensive to conduct, and proper interpretation of the results is often very critical to decision making. It has been estimated that nearly 50% of all experiments are of the "nonreplicated" type and many of these are not being analyzed statistically because researchers are not aware of existing statistical methods which can be used. This book tries to provide researchers with statistical methods appropriate for nonreplicated experiments as well as some ways to make the required statistical computation feasible using existing statistical software.

Many experiments are very expensive to conduct, and requiring independent replications of all of the treatment combinations can be overly burdensome to experimenters. For example, suppose an automobile engineer wishes to determine optimal locations for seatbelt anchors. Locations must be determined so that safety and comfort are assured for drivers and occupants of many different sizes. Running an experiment which uses all possible combinations, the influential treatment factors may require 200 to 300 different runs for a single replication. Each run may cost several thousand dollars because simulated automobile crashes are expensive to conduct. Obviously even a single replication of such an experiment would be much too expensive to run.

The basic purpose of this book is to introduce several techniques and methods for analyzing experiments in which there are no independent replications of the treatment combinations being studied. Occasionally there have been clients who have several measurements on each experimental unit which they had planned to analyze as though these measurements represented independent replications of their treatment combinations. Some of the techniques presented in this book have helped to salvage information from these kinds of experiments. Many of the techniques discussed in this book are not currently available in any other books.

Users of this book will learn the following:

1. How to recognize whether replications are independent replications or dependent replications

2. How to test for interaction in nonreplicated experiments

3. How to obtain reasonable estimates of the experimental error variance in nonreplicated experiments

4. How to determine whether Tukey's model or Mandel's model can be used to model the data collected in an experiment

5. When to use a multiplicative interaction model and how to use it

6. How to use existing statistical software to fit these models

7. How to determine which treatment combinations are the primary causes of significant interactions

8. How to construct interaction plots and how to use them to interpret data

9. Simplified procedures for constructing half-normal plots

10. How to use half-normal plots to your advantage

11. How to use $2^n$ factorial experiments for exploratory purposes

12. How to use blocking to conduct $2^n$ factorial experiments more efficiently

13. How to use fractional replications of factorial experiments for exploratory purposes

14. How to use polynomial models to model certain kinds of nonreplicated experiments

15. How to use quadratic response surface models and contour plots to advantage when developing new products or improving old products

The approach used in this book is similar to that used in the first volume. That is, each topic is covered from a practical viewpoint, emphasizing the implementation of the methods much more than the theory behind the methods. Many real-world examples are used to illustrate the techniques introduced. Formulas are included for those readers who would like to program the techniques on their personal computers.

The book is intended for everyone who analyzes data. The reader should have a knowledge of analysis of variance techniques as well as basic statistical ideas. Although a knowledge of the contents of *Analysis of Messy Data Vol. 1—Designed Experiments* would be useful, such knowledge is not required.

The book contains several tables that are not available in many other books and examples that will help readers recognize the need for a particular method and show how the method should be correctly used for their own situations.

We would like to express our appreciation to Linda Kaufholz who did the initial typing of this manuscript, and we especially appreciate her willingness to learn word processing while working on this book. Her expertise played a major role in the final project. We would also like to thank Retha Parker for using her talents in the revisions of the original manuscript.

# 1

# Analyzing Two-Way Treatment Structures with One Observation per Treatment Combination

**CHAPTER OUTLINE**

$\mathbf{M}$any experiments are very expensive to conduct so that experimenters are often forced to limit the number of treatment combinations that can be studied in order to have adequate resources available to replicate the treatment combinations under study. The replication of treatment combinations is necessary in order to be able to have an independent estimate of the experimental error variance, which is denoted by $\sigma^2$ in this book. Having a good estimate of $\sigma^2$ is very important when the major objective of the experiment is confirmatory. However, if the major objective of the experiment is exploratory, it is often more desirable to study many different treatment combinations, each performed once, rather than a few treatment combinations each replicated many times. This book is devoted to methods that can help to extract the relevant information in experiments that are not replicated.

The first three chapters of this book are specifically devoted to analyzing two-way treatment structure experiments that have not been replicated. That is, there is only one independent observation for each treatment combination. If the cost of conducting the experiment is relatively insignificant, one cannot generally recommend designing experiments in this fashion. That is, if cost is not a factor, then one can almost always obtain better information from replicated experiments than from nonreplicated experiments.

## 1.1 INTRODUCTION

Suppose an experimenter wants to study the effect of temperature and humidity on the growth of a particular variety of sorghum when there are 12 growth chambers within which both the level of humidity and temperature can be controlled. The experimenter decides to study the effects of these two factors on sorghum growth by studying the 12 Temperature*Humidity combinations that are generated by considering all possible combinations of three temperature levels with four humidity levels. Typically, the experimenter places several plants of the specified variety within each growth chamber. Some researchers incorrectly analyze the observations on these plants as though they are independent replications of the 12 Temperature*Humidity treatment combinations. However, such observations are merely subsamples or repeated measures on the growth chambers rather than independent replications of the 12 Temperature*Humidity treatment combinations. The experimental units for this experiment are the 12 growth chambers; the fact that there are several plants within each growth chamber merely means that each growth chamber is being measured several times. With 12 treatment combinations and 12 experimental units, the experimenter has only one observation for each treatment combination and the usual methods for statistical analysis of the observed results do not apply because there are no independent replications from which to estimate $\sigma^2$. Thus, the experimenter is faced with several alternatives, none being very desirable. The alternatives are:

1. Decrease the number of Temperature∗Humidity treatment combinations to be studied. For instance, if the number of combinations is reduced to 6, each combination could be assigned to two growth chambers resulting in two independent replications of each treatment combination.

2. Plan to repeat the experiment again at a later time using the same growth chambers, but rerandomizing the Temperature∗Humidity combinations that are to be assigned to these growth chambers. This may be a viable alternative for fast-growing plants, but may not be realistic for slow-growing plants.

3. Conduct the experiment as planned, and use some of the analysis techniques described in this book.

This experiment illustrates a situation where it is possible to replicate the experiment even though the experiment may not actually get replicated. There are other situations where replicating the experiment is impossible. One such case is where an experimenter wants to compare the protein content of several varieties of wheat grown at many different locations. Such an experiment is impossible to replicate, since locations cannot be replicated. The methods described in this book will often enable experimenters to obtain usable information from experiments that are only replicated once.

Many experimenters have been observed to conduct experiments involving two-way treatment structures where the resulting data provides only one observation per treatment combination. These single-observation experiments often occur by accident. That is, the experimenter thought he or she was replicating the experiment while, in reality, the so-called replicates were really subsamples. Many experimenters have difficulty seeing the difference between true independent replications and subsampling. Those readers who have difficulty seeing the difference are advised to read Chapters 4 and 5 of Milliken and Johnson (1984). These chapters discuss split-plot and/or repeated-measures experiments. Subsampling is similar to a split-plot experiment except that no new treatments are applied to the subplot experimental units. An example involving only one independent replication per treatment combination is described in the next section.

**1.2 AN EXAMPLE**

Next, we give a numerical example. The example, complete with data, is used in subsequent sections to demonstrate some of the techniques for analyzing nonreplicated two-way experiments.

**EXAMPLE 1.1: Growth Rate of Sorghum Plants** ⎯⎯⎯⎯⎯⎯⎯⎯⎯⎯⎯⎯⎯

An experimenter has 20 growth chambers and conducts an experiment to study the effects of five temperature levels combined with each of four

## Table 1.1   Mean Height of 10 Sorghum Plants

| | *Humidity, %* | | | |
|---|---|---|---|---|
| *Temperature, °F* | *20* | *40* | *60* | *80* |
| 50 | 12.3 | 19.6 | 25.7 | 30.4 |
| 60 | 13.7 | 16.9 | 27.0 | 31.5 |
| 70 | 17.8 | 20.0 | 26.3 | 35.9 |
| 80 | 12.1 | 17.4 | 36.9 | 43.4 |
| 90 | 6.9 | 18.8 | 35.0 | 53.0 |

humidity levels on the growth rate of sorghum plants. The experimenter places 10 sorghum plants of the same species in each of the 20 growth chambers and assigns Temperature∗Humidity combinations randomly to the 20 chambers. The data given in Table 1.1 represent average heights in centimeters of the 10 plants from each growth chamber. These heights were measured after growing the plants for 4 weeks in the growth chambers. The average height of the 10 plants is used as the response because the experimental units for the Temperature∗Humidity treatment combinations are the growth chambers. There is only one independent replication of each growth chamber. The variability existing between the 10 plants within a growth chamber does not measure the variability between growth chambers. That is, variability between plants growing within the same growth chamber may be much different than variability between plants growing in different growth chambers even if different growth chambers had been assigned the same Temperature∗Humidity combinations.

As noted for split-plot and repeated-measures experiments, the within-growth-chamber variability is not an appropriate measure of variability with which to compare the effects of the treatments observed on the growth chambers. Several methods for analyzing this type of data are presented in the following sections.

## 1.3 MODEL DEFINITION AND PARAMETER ESTIMATION

A means model for experiments such as the one described in Example 1.1 is given by

$$y_{ij} = \mu_{ij} + \epsilon_{ij} \qquad i = 1, 2, \ldots, t, \quad j = 1, 2, \ldots, b \qquad (1.3.1)$$

where

$$\epsilon_{ij} \sim \text{i.i.d. } N(0, \sigma^2) \qquad i = 1, 2, \ldots, t, \quad j = 1, 2, \ldots, b$$

where i.i.d. means independently and identically distributed. In this model there are $bt$ parameters and $bt$ observations. The best estimate of $\mu_{ij}$ is $\hat{\mu}_{ij} = y_{ij}$,   $i = 1, 2, \ldots t$,   $j = 1, 2, \ldots b$. However, no estimate of $\sigma^2$ is available, unless one is able to make some assumptions about the $\mu_{ij}$'s. One assumption made by many statistical analysts is that the two sets of treatments do not interact. This is equivalent to assuming $\mu_{ij} - \mu_{i'j} - \mu_{ij'} + \mu_{i'j'} = 0$ for all possible values of $i, i', j$, and $j'$. As stated in Chapter 7 of Milliken and Johnson (1984), it is also true that there is no interaction between the levels of the two sets of treatment combinations in an experiment if and only if there exist parameters $\mu$, $\tau_1$, $\tau_2$, $\ldots$, $\tau_t$, $\beta_1$, $\beta_2$, $\ldots$, $\beta_b$ such that

$$\mu_{ij} = \mu + \tau_i + \beta_j \qquad i = 1, 2, \ldots, t, \quad j = 1, 2, \ldots, b$$

If it is true that there is no interaction between the levels of the treatments, the best estimate of $\sigma^2$ is

$$\hat{\sigma}^2 = \sum_{i=1}^{t} \sum_{j=1}^{b} (y_{ij} - \bar{y}_{i.} - \bar{y}_{.j} + \bar{y}_{..})^2 / (b-1)(t-1)$$

The assumption of no interaction should not be made without some justification for it being true. However, we have found many experimenters more than willing to assume their treatments do not interact, especially when such an assumption enables them to calculate some test statistics. In fact, some experimenters do not put interaction terms in their models because they think they are not interested in the interaction. In reality, they have no choice but to be interested in interaction, if it exists; hence it is important to determine whether or not interaction exists.

## 1.4 WHAT CAN BE DONE?

What can an experimenter do when interaction is suspected to be present in the data or when the experimenter wants to test for it? Several methods are available to help answer this question. Unfortunately, no method is best for all situations, and most of the available methods assume that the $\mu_{ij}$'s can be described by some sort of model other than a simple additive model. Because of this assumption, most available tests for interaction are good for certain types of interaction but not for all types.

Each of the remaining sections of this chapter presents a test for interaction for the two-way treatment structure with one observation per treatment combination.

## 1.5 A HEURISTIC TEST FOR INTERACTION

The test being considered in this section was presented by Milliken and Rasmuson (1977). One advantage this test has over the remaining tests presented in this chapter is that it does not require any assumptions about the form of the interaction; however, a disadvantage is that there are some forms of interaction that the method cannot detect.

The procedure can be described as follows:

1. Partition the observations by the levels of factor $T$ into the $t$ sets,

$$\{ y_{11}, y_{12}, \ldots, y_{1b} \}, \{ y_{21}, y_{22}, \ldots, y_{2b} \}, \ldots, \{ y_{t1}, y_{t2}, \ldots, y_{tb} \}$$

2. Determine the variance of the observations in each set. That is, let $v_i^2 = \sum_{j=1}^{b} (y_{ij} - \bar{y}_{i\bullet})^2 / (b - 1)$, $i = 1, 2, \ldots, t$. Next note that if there is no interaction, then $\mu_{ij} = \mu + \tau_i + \beta_j$ and each $v_i^2$ is an unbiased estimate of

$$\sigma^2 + \frac{1}{b - 1} \sum_{j=1}^{b} (\beta_j - \bar{\beta}_\bullet)^2$$

However, when there is interaction, then $\mu_{ij} = \mu + \tau_i + \beta_j + \gamma_{ij}$, and each $v_i^2$ is an unbiased estimate of

$$\sigma^2 + \frac{1}{b - 1} \sum_{j=1}^{b} (\beta_j - \bar{\beta}_\bullet + \gamma_{ij} - \gamma_{i\bullet})^2 = \delta_i^2 \quad \text{(say)}$$

Hence, if one tests $H_0$: $\delta_1^2 = \delta_2^2 = \cdots = \delta_t^2$ and rejects, then one can conclude that there is interaction in the data.

3. To test $H_0$: $\delta_1^2 = \delta_2^2 = \cdots = \delta_t^2$, Milliken and Rasmuson recommend using any of the tests for homogeneity of variance given in Chapter 2 of Milliken and Johnson (1984).

For the interested reader, the $v_i^2$'s are multiples of noncentral chi-square random variables, rather than central chi-square random variables as required for the tests in Milliken and Johnson. Thus, the homogeneity tests are only approximate for the situation described here.

One unfortunate aspect of the above test is that even when one accepts $H_0$, one is still not able to conclude that the data are additive. This is true since it is possible for there to be interaction in the data and still have all the $\delta_i^2$ equal. Thus, if $H_0$ is rejected, there is interaction in the data, but if $H_0$ is accepted, one cannot guarantee no interaction. When $H_0$ is accepted, one could partition the data according to the levels of $B$, and test for equality of variances in the $b$ sets of $t$ observations each, using the procedure described above.

Unfortunately, even if this hypothesis is also accepted, one still cannot conclude that the data are additive. This can be illustrated by examining the following set of true cell means.

| $\mu_{ij}$ | | B | | Row Variance |
|:---:|:---:|:---:|:---:|:---:|
| | 1 | 2 | 3 | |
| 1 | 8 | 6 | 4 | 4 |
| T  2 | 4 | 8 | 6 | 4 |
| 3 | 6 | 4 | 8 | 4 |
| Column Variance | 4 | 4 | 4 | |

For the above data, all row variances are equal and all column variances are equal, and thus the heuristic test for interaction will not detect the interaction which actually exists between the two sets of treatment factors.

As an example, consider the data in Table 1.1. The column variances are 15.268, 1.828, 28.407, and 88.763, respectively. Hartley's $F$-max statistic is

$$F_{max} = \frac{88.763}{1.828} = 48.56$$

which is significant at, approximately, the 1% level $\left[F_{max,4,4}(.01) = 49.0\right]$. Critical points for this statistic are tabled in Milliken and Johnson (1984). Bartlett's test statistic is equal to

$$\chi_c^2 = 10.48$$

which is also almost significant at the 1% level ($\chi_{.01,3}^2 = 11.345$).

Thus, one would conclude that interaction exists between the levels of the two sets of treatments in these data.

Tukey (1949) was the first writer to propose a test for interaction in the two-way treatment structure experiment with one observation per treatment combination. Although Tukey did not consider any particular model when he proposed the test, other authors, Ward and Dick (1952), Scheffé (1959), and Graybill (1961, 1977), showed that the test is most easily motivated by assuming that the cell means can be expressed as

**1.6 TUKEY'S SINGLE-DEGREE-OF-FREEDOM TEST FOR NONADDITIVITY**

$$\mu_{ij} = \mu + \tau_i + \beta_j + \lambda\tau_i\beta_j$$
$$i = 1,2,\ldots,t, \quad j = 1,2,\ldots,b \qquad (1.6.1)$$

That is, it is assumed that the interaction term $\gamma_{ij}$ in the usual effects model

$$\mu_{ij} = \mu + \tau_i + \beta_j + \gamma_{ij}$$
$$i = 1,2,\ldots,t, \quad j = 1,2,\ldots,b \qquad (1.6.2)$$

is a scalar multiple of the product of the row and column main effects; i.e., $\gamma_{ij} = \lambda \tau_i \beta_j$, $i = 1, 2, \ldots, t$, $j = 1, 2, \ldots, b$. A test for interaction is made by testing $H_0$: $\lambda = 0$ versus $H_0$: $\lambda \neq 0$.

Tukey proposed using the sum of squares

$$\text{SSN} = \frac{\left[ \sum_{ij} (\bar{y}_{i.} - \bar{y}_{..})(\bar{y}_{.j} - \bar{y}_{..})(y_{ij} - \bar{y}_{i.} - \bar{y}_{.j} + \bar{y}_{..}) \right]^2}{\sum_i (\bar{y}_{i.} - \bar{y}_{..})^2 \sum_j (\bar{y}_{.j} - \bar{y}_{..})^2} \tag{1.6.3}$$

as a measure of nonadditivity. When $H_0$: $\lambda = 0$ is true, $\text{SSN}/\sigma^2$ has a sampling distribution that is chi-square with 1 degree of freedom. The residual sum of squares after fitting the interaction term is

$$\text{SSR} = \sum_{ij} (y_{ij} - \bar{y}_{i.} - \bar{y}_{.j} + \bar{y}_{..})^2 - \text{SSN} \tag{1.6.4}$$

When $H_0$: $\lambda = 0$ is true, $\text{SSR}/\sigma^2$ has a chi-square distribution with $(b-1)(t-1) - 1$ degrees of freedom, and SSR is distributed independently of SSN. Tukey's single-degree-of-freedom test for nonadditivity is: Reject $H_0$: $\lambda = 0$ if

$$F_c = \text{SSN}/[\text{SSR}/(bt - b - t)] > F_{\alpha, 1, bt - b - t}$$

Once again, if we fail to reject $H_0$, we can conclude there is no interaction of the form $\lambda \tau_i \beta_j$, but still cannot really conclude the data are additive.

**1.6.1
Computing
Tukey's Test
Using
Statistical
Software**

Tukey's test statistic for nonadditivity can be obtained using many existing statistical packages. The procedure requires two steps. The first step consists of fitting the additive model

$$y_{ij} = \mu + \tau_i + \beta_j + \epsilon_{ij} \tag{1.6.5}$$

and selecting the solution of the normal equations [see Milliken and Johnson (1984), Chapter 6] that satisfies $\sum \hat{\tau}_i = 0$ and $\sum \hat{\beta}_j = 0$. In this case, $\hat{\tau}_i$ and $\hat{\beta}_j$ are given by $\hat{\tau}_i = \bar{y}_{i.} - \bar{y}_{..}$, $i = 1, 2, \ldots, t$ and $\hat{\beta}_j = \bar{y}_{.j} - \bar{y}_{..}$, $j = 1, 2, \ldots, b$. Then equation (1.6.3) simplifies to

$$\text{SSN} = \left[ \sum_{ij} \hat{\tau}_i \hat{\beta}_j z_{ij} \right]^2 / \left[ \sum_{ij} (\hat{\tau}_i \hat{\beta}_j)^2 \right]$$

where

$$z_{ij} = y_{ij} - \bar{y}_{i.} - \bar{y}_{.j} + \bar{y}_{..}$$

Secondly, one may obtain SSN using a statistical computing package by fitting the model

$$z_{ij} = \lambda \hat{\tau}_i \hat{\beta}_j + e_{ij}$$

Fitting this model is similar to fitting a simple linear regression model without an intercept. Hence,

$$\hat{\lambda} = \left[ \sum_{ij} \hat{\tau}_i \hat{\beta}_j z_{ij} \right] \bigg/ \sqrt{ \sum_{ij} \hat{\tau}_i^2 \hat{\beta}_j^2 }$$

and the sum of squares due to regression is

$$\left[ \sum_{ij} \hat{\tau}_i \hat{\beta}_j z_{ij} \right]^2 \bigg/ \left[ \sum_{ij} \hat{\tau}_i^2 \hat{\beta}_j^2 \right] = \text{SSN}$$

The quantity SSR needed for the $F$-test is obtained by subtracting SSN from the residual sum of squares obtained from fitting the additive model (1.6.5). It is interesting to note that if one replaces $z_{ij}$ in the above two equations by $y_{ij}$, the results are the same.

Tukey's test can be obtained with SAS® by following the steps below:

Step 1: Fit model (1.6.5) by using

**1.6.2
Computing
Tukey's Test
Using SAS®**

PROC GLM; CLASSES T B;

MODEL Y = T B/SOLUTION;

The solution vector given by SAS® satisfies $\hat{\tau}_t = 0$ and $\hat{\beta}_b = 0$ rather than $\sum \hat{\tau}_i = 0$ and $\sum \hat{\beta}_j = 0$. A new solution vector $[\hat{\mu}^*, \hat{\tau}_1^*, \ldots, \hat{\tau}_t^*, \hat{\beta}_1^*, \ldots, \hat{\beta}_b^*]'$, which satisfies $\sum \hat{\tau}_i^* = 0$ and $\sum \hat{\beta}_j^* = 0$ can be obtained from the SAS® solution by letting:

$$\hat{\mu}^* = \hat{\mu} + \bar{\hat{\tau}}_{\bullet} + \bar{\hat{\beta}}_{\bullet}$$
$$\hat{\tau}_i^* = \hat{\tau}_i - \bar{\hat{\tau}}_{\bullet}, \qquad i = 1, 2, \ldots, t$$
$$\hat{\beta}_j^* = \hat{\beta}_j - \bar{\hat{\beta}}_{\bullet}, \qquad j = 1, 2, \ldots, b \qquad (1.6.6)$$

Step 2: Construct a new set of data by adding both $\hat{\tau}_i^*$ and $\hat{\beta}_j^*$ to the data card, which contains the treatment combination $T_i$ and $B_j$, and then fit the model,

$$y_{ij} = \mu + \tau_i + \beta_j + \lambda \hat{\tau}_i^* \hat{\beta}_j^* + \epsilon_{ij}$$

by using

PROC GLM; CLASSES T B;

MODEL Y = T B THAT*BHAT/SOLUTION;

The $F$-test given by either the Type I or Type III analysis that corresponds to the row labeled by THAT∗BHAT is Tukey's single-degree-of-freedom test for nonadditivity.

**1.6.3**
**An Example**
**Illustrating**
**Tukey's Test**

Consider the data in Table 1.1. The data were first analyzed with SAS® GLM using the statements:

PROC GLM;

CLASSES TEMP HUMIDITY;

MODEL HEIGHT = TEMP HUMIDITY / SOLUTION;

The results obtained from the SOLUTION option are shown in Table 1.2.

The estimates given in Table 1.2 are substituted into equation (1.6.6) to find the sum-to-zero solutions. First note that

$$\bar{\hat{\tau}}_. = -3.395 \text{ and } \bar{\hat{\beta}}_. = -13.810$$

Then

$$\hat{\mu}^* = 42.235 + (-3.395) + (-13.810) = 25.03$$
$$\hat{\tau}_1^* = -6.425 - (-3.395) = -3.030$$
$$\hat{\tau}_2^* = -6.150 - (-3.395) = -2.755$$
$$\hat{\tau}_3^* = -3.425 - (-3.395) = -.030$$
$$\hat{\tau}_4^* = -.975 - (-3.395) = 2.420$$
$$\hat{\tau}_5^* = .000 - (-3.395) = 3.395$$
$$\hat{\beta}_1^* = -26.280 - (-13.81) = -12.47$$
$$\hat{\beta}_2^* = -20.30 - (-13.81) = -6.49$$
$$\hat{\beta}_3^* = -8.66 - (-13.81) = 5.15$$
$$\hat{\beta}_4^* = .00 - (-13.81) = 13.81$$

The data set to be analyzed next is given in Table 1.3 and is equivalent to the original data set augmented by the $\tau_i^*$ and $\beta_j^*$.

The data in Table 1.3 are analyzed with SAS® GLM by using the commands:

PROC GLM;

CLASSES TEMP HUMIDITY;

MODEL HEIGHT = TEMP HUMIDITY THAT∗HHAT/SOLUTION;

The Type I and Type III analysis of variance tables generated by SAS® GLM are given in Table 1.4 and the results from the SOLUTION option are given in Table 1.5.

**Table 1.2  Results from SOLUTION Option of SAS® GLM**

| Parameter | | Estimate | $T$ for $H_0$: Parameter $= 0$ | $PR > |T|$ | STD Error of Estimate |
|---|---|---|---|---|---|
| Intercept | | 42.23500000 B | 11.56 | 0.0001 | 3.65352341 |
| Temp | 50 | $-6.42500000$ B | $-1.57$ | 0.1417 | 4.08476336 |
| | 60 | $-6.15000000$ B | $-1.51$ | 0.1580 | 4.08476336 |
| | 70 | $-3.42500000$ B | $-0.84$ | 0.4181 | 4.08476336 |
| | 80 | $-0.97500000$ B | $-0.24$ | 0.8154 | 4.08476336 |
| | 90 | 0.00000000 B | . | . | . |
| Humidity | 20 | $-26.28000000$ B | $-7.19$ | 0.0001 | 3.65352341 |
| | 40 | $-20.30000000$ B | $-5.56$ | 0.0001 | 3.65352341 |
| | 60 | $-8.66000000$ B | $-2.37$ | 0.0354 | 3.65352341 |
| | 80 | 0.00000000 B | . | . | . |

**Table 1.3  Data for Tukey's Model**

| Temp | Humidity | Y | THAT | HHAT |
|---|---|---|---|---|
| 50 | 20 | 12.3 | $-3.030$ | $-12.47$ |
| 50 | 40 | 19.6 | $-3.030$ | $-6.49$ |
| 50 | 60 | 25.7 | $-3.030$ | 5.15 |
| 50 | 80 | 30.4 | $-3.030$ | 13.81 |
| 60 | 20 | 13.7 | $-2.755$ | $-12.47$ |
| 60 | 40 | 16.9 | $-2.755$ | $-6.49$ |
| 60 | 60 | 27.0 | $-2.755$ | 5.15 |
| 60 | 80 | 31.5 | $-2.755$ | 13.81 |
| 70 | 20 | 17.8 | $-0.030$ | $-12.47$ |
| 70 | 40 | 20.0 | $-0.030$ | $-6.49$ |
| 70 | 60 | 26.3 | $-0.030$ | 5.15 |
| 70 | 80 | 35.9 | $-0.030$ | 13.81 |
| 80 | 20 | 12.1 | 2.420 | $-12.47$ |
| 80 | 40 | 17.4 | 2.420 | $-6.49$ |
| 80 | 60 | 36.9 | 2.420 | 5.15 |
| 80 | 80 | 43.4 | 2.420 | 13.81 |
| 90 | 20 | 6.9 | 3.395 | $-12.47$ |
| 90 | 40 | 18.8 | 3.395 | $-6.49$ |
| 90 | 60 | 35.0 | 3.395 | 5.15 |
| 90 | 80 | 53.0 | 3.395 | 13.81 |

**Table 1.4   Type I and Type III Analysis of Variance Tables Generated by SAS® GLM**

| Source | DF | Sum of squares | | Mean square |
|---|---|---|---|---|
| Model | 8 | 2499.56700865 | | 312.44587608 |
| Error | 11 | 111.79499135 | | 10.16318103 |
| Corrected Total | 19 | 2611.36200000 | | |
| Model F = | 30.74 | | | PR > F = 0.0001 |

| R-square | C.V. | Root MSE | | Y mean |
|---|---|---|---|---|
| 0.957189 | 12.7366 | 3.18797444 | | 25.03000000 |

| Source | DF | Type I SS | F value | PR > F |
|---|---|---|---|---|
| Temp | 4 | 136.61700000 | 3.36 | 0.0498 |
| Humidity | 3 | 2074.29800000 | 68.03 | 0.0001 |
| THAT*HHAT | 1 | 288.65200865 | 28.40 | 0.0002 |

| Source | DF | Type III SS | F value | PR > F |
|---|---|---|---|---|
| Temp | 4 | 136.61700000 | 3.36 | 0.0498 |
| Humidity | 3 | 2074.29800000 | 68.03 | 0.0001 |
| THAT*HHAT | 1 | 288.65200865 | 28.40 | 0.0002 |

From Table 1.4 we see that Tukey's single-degree-of-freedom test for nonadditivity is $F = 28.40$ and $H_0$: $\lambda = 0$ is rejected at the $\alpha = .0002$ significance level. Also the value of SSN in (1.6.3) is 288.652 and the value of SSR in (1.6.4) is 111.795 and has 11 degrees of freedom. From the last row of Table 1.5, the estimate of $\lambda$ in the model (1.6.1) is $\hat{\lambda} = .14273$.

**Table 1.5   Results from SOLUTION Option of SAS® GLM with Tukey's Model**

| Parameter | | Estimate | T for $H_0$: Parameter = 0 | PR > \|T\| | STD Error of Estimate |
|---|---|---|---|---|---|
| Intercept | | 42.23500000 B | 20.95 | 0.0001 | 2.01625207 |
| Temp | 50 | − 6.42500000 B | − 2.85 | 0.0158 | 2.25423834 |
| | 60 | − 6.15000000 B | − 2.73 | 0.0196 | 2.25423834 |
| | 70 | − 3.42500000 B | − 1.52 | 0.1569 | 2.25423834 |
| | 80 | − 0.97500000 B | − 0.43 | 0.6737 | 2.25423834 |
| | 90 | 0.00000000 B | . | . | . |
| Humidity | 20 | − 26.28000000 B | − 13.03 | 0.0001 | 2.01625207 |
| | 40 | − 20.30000000 B | − 10.07 | 0.0001 | 2.01625207 |
| | 60 | − 8.66000000 B | − 4.30 | 0.0013 | 2.01625207 |
| | 80 | 0.00000000 B | . | . | . |
| THAT*HHAT | | 0.14272970 | 5.33 | 0.0002 | 0.02678193 |

In this section two types of interaction plots are introduced. These plots are useful for helping one choose an appropriate model for two-way treatment structure experiments with interaction.

To demonstrate interaction plots, consider a set of true cell mean parameters satisfying $\mu_{ij} = \mu + \tau_i + \beta_j$ where $\mu = 29$, $\tau_1 = -5$, $\tau_2 = 2$, $\tau_3 = 4$, $\tau_4 = 0$, $\tau_5 = -1$, $\beta_1 = -5$, $\beta_2 = 2$, $\beta_3 = -1$, and $\beta_4 = 4$. Note that, without any loss of generality, the $\tau$'s and $\beta$'s have been chosen so that they satisfy $\sum \tau_i = \sum \beta_j = 0$. The cell means for this hypothetical model are given in Table 1.6.

**1.7.1**
**Interaction**
**Plots**
**for Additive**
**Cell Mean**
**Parameters**

The first type of interaction plot to be considered, called a Type I interaction plot, is illustrated in Figure 1.1. This type of plot is constructed by plotting $\mu_{ij}$ against $i$ (or $j$) for each value of $j$ (or $i$). This is the kind of plot introduced in most textbooks.

When the cell mean parameters are from an additive model the plot will always consist of line segments which are parallel. Data with interaction will have at least one line segment which is not parallel to the others.

Another type of interaction plot, called a Type II interaction plot, is constructed by plotting $\mu_{ij}$ against $\tau_i$ (or $\beta_j$) for each value of $j$ (or $i$). Such a plot is illustrated in Figure 1.2. Note that for data from an additive model, a Type II interaction plot always consists of parallel lines rather than parallel line segments.

Next Type I and Type II interaction plots for a set of cell mean parameters satisfying Tukey's model are illustrated.

Let $\mu_{ij} = \mu + \tau_i + \beta_j + \lambda \tau_i \beta_j$ with $\mu$, $\tau$'s, and $\beta$'s defined as above and with $\lambda = 1$. The cell mean parameters for this case are given in Table 1.7.

**1.7.2**
**Interaction**
**Plots for Cell**
**Means**
**Satisfying**
**Tukey's Model**

A Type I interaction plot for these cell means is given in Figure 1.3, and a Type II interaction plot is given in Figure 1.4. Examination of Figure 1.3 reveals little more other than the cell mean parameters are from a nonadditive model. Examination of Figure 1.4, however, reveals

### Table 1.6  Cell Mean Parameters of an Additive Model

| $\mu_{ij}$ | $B_1$ | $B_2$ | $B_3$ | $B_4$ |
|---|---|---|---|---|
| $T_1$ | 19 | 26 | 23 | 28 |
| $T_2$ | 26 | 33 | 30 | 35 |
| $T_3$ | 28 | 35 | 32 | 37 |
| $T_4$ | 24 | 31 | 28 | 33 |
| $T_5$ | 23 | 30 | 27 | 32 |

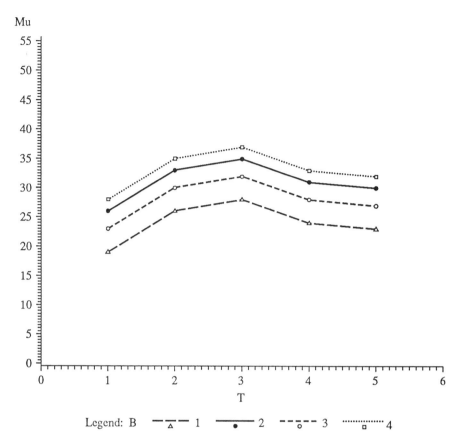

**Figure 1.1   A Type I Interaction Plot of Additive Cell Mean Parameters.**

that cell means from a model satisfying Tukey's model provide a very distinctive interaction plot. The plot consists of several straight lines, all intersecting in a single point. This will always be the case for data which can be modeled by Tukey's model, regardless of whether one plots $\mu_{ij}$ against $\tau_i$ for each $j$, or plots $\mu_{ij}$ against $\beta_j$ for each $i$.

**Table 1.7   Cell Mean Parameters for Tukey's Model**

| $\mu_{ij}$ | $B_1$ | $B_2$ | $B_3$ | $B_4$ |
|---|---|---|---|---|
| $T_1$ | 44 | 16 | 28 | 8 |
| $T_2$ | 16 | 37 | 28 | 43 |
| $T_3$ | 8 | 43 | 28 | 53 |
| $T_4$ | 24 | 31 | 28 | 33 |
| $T_5$ | 28 | 28 | 28 | 28 |

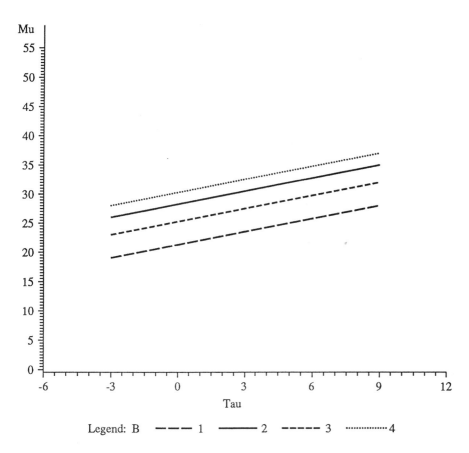

**Figure 1.2   A Type II Interaction Plot of Additive Cell Mean Parameters.**

A Type II interaction plot of the data in Table 1.1 is given in Figure 1.5. The plot in Figure 1.5 uses temperature as a baseline. The numbers on the plot indicate the humidity level.

**1.7.3 Type II Interaction Plots for the Data in Example 1.1**

Figure 1.5 also contains estimated regression lines obtained by fitting Tukey's model (1.6.1) to the data in Table 1.1. Points lying on the estimated regression lines can be obtained from

$$\hat{\mu}_{ij} = \hat{\mu} + \hat{\tau}_i + \hat{\beta}_j + \hat{\lambda}\hat{\tau}_i\hat{\beta}_j$$

They could also be obtained by using a P option on the MODEL statement in SAS® GLM.

To illustrate the computation of the predicted values consider the point corresponding to TEMPERATURE = 50 and HUMIDITY = 20. For this combination THAT = −3.03 and HHAT = −12.47 (see Table 1.3). Also $\hat{\mu} = \bar{y}_{..}$ = 25.03 and $\hat{\lambda}$ = .14273 from Table 1.5. Thus

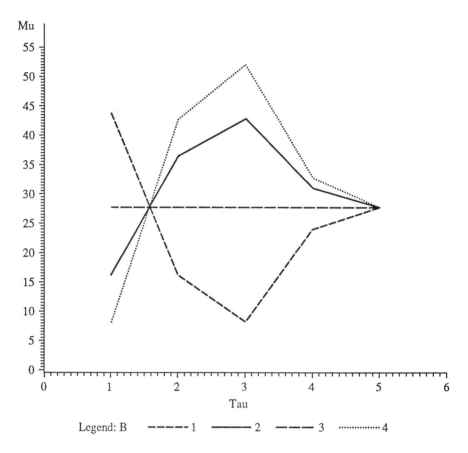

**Figure 1.3   A Type I Interaction Plot for Tukey's Model.**

$$\hat{\mu}_{11} = 25.03 + (-3.03) + (-12.47)$$
$$+ (.14273)(-3.03)(-12.47) = 14.92$$

This is the point labeled A in Figure 1.5. As a second illustration, consider the point corresponding to TEMPERATURE = 90 and HUMIDITY = 60. For this combination THAT = 3.395 and HHAT = 5.15. Hence

$$\hat{\mu}_{53} = 25.03 + 3.395 + 5.15$$
$$+ (.14273)(3.395)(5.15) = 36.07$$

This is the point labeled B in Figure 1.5. In a similar manner other points on these prediction lines can be obtained through which the lines can be drawn.

A second Type II interaction plot for these data and Tukey's model is given in Figure 1.6. The plot in Figure 1.6 uses Humidity as the baseline.

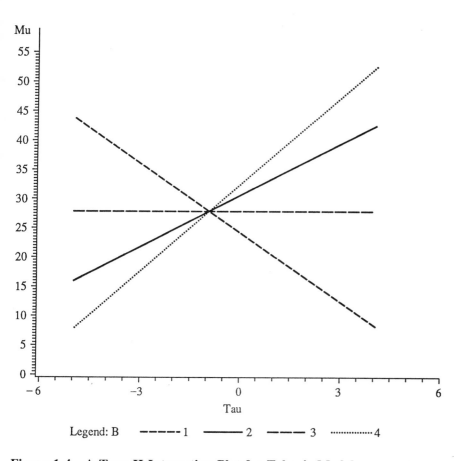

**Figure 1.4   A Type II Interaction Plot for Tukey's Model.**

Mandel (1959) generalized Tukey's model to

$$\mu_{ij} = \mu + \tau_i + \beta_j + \lambda\alpha_i\beta_j$$

$$i = 1, 2, \ldots, t, \quad j = 1, 2, \ldots, b \qquad (1.8.1)$$

An alternative form of model (1.8.1) is obtained by letting $\delta_i = 1 + \lambda\alpha_i$, then (1.8.1) becomes

$$\mu_{ij} = \mu + \tau_i + \delta_i\beta_j$$

$$i = 1, 2, \ldots, t, \quad j = 1, 2, \ldots, b \qquad (1.8.2)$$

If $\alpha_1 = \alpha_2 = \cdots = \alpha_t$ in (1.8.1) or equivalently if $\delta_1 = \delta_2 = \cdots = \delta_t$ in (1.8.2), then Mandel's model reduces to an additive model. Hence, a test for interaction can be obtained by testing $H_0$: $\alpha_1 = \alpha_2 = \cdots = \alpha_t$. Once again, if $H_0$ is accepted, one cannot necessarily conclude

**1.8 MANDEL'S BUNDLE-OF-STRAIGHT-LINES MODEL**

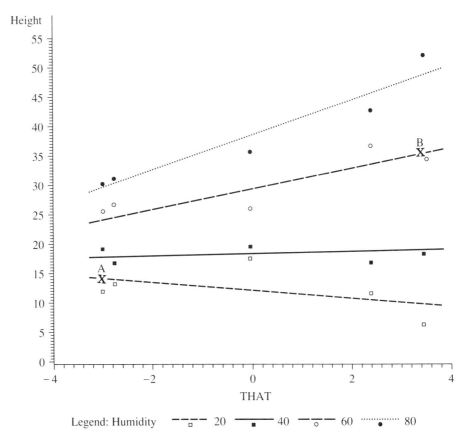

**Figure 1.5    A Type II Interaction Plot for the Data in Table 1.1 Using Temperature as a Baseline. Lines Are Those Estimated by Tukey's Model.**

that there is no interaction, but only that there is no interaction of the form $\alpha_i \beta_j$.

The cell means given in Table 1.8 are obtained from Mandel's model by letting $\mu$, the $\tau$'s, and the $\beta$'s be as defined in Section 1.7, and by letting $\lambda = 1, \alpha_1 = 1, \alpha_2 = 2, \alpha_3 = 4, \alpha_4 = 0$, and $\alpha_5 = -2$.

**Table 1.8    Cell Means for Mandel's Model**

| $\mu_{ij}$ | $B_1$ | $B_2$ | $B_3$ | $B_4$ |
|---|---|---|---|---|
| $T_1$ | 14 | 28 | 22 | 32 |
| $T_2$ | 16 | 37 | 28 | 43 |
| $T_3$ | 8 | 43 | 28 | 53 |
| $T_4$ | 24 | 31 | 28 | 33 |
| $T_5$ | 33 | 26 | 29 | 24 |

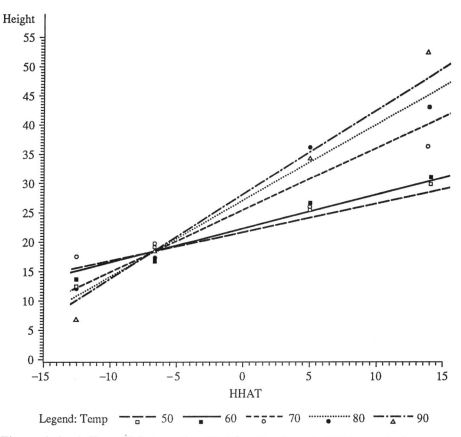

**Figure 1.6  A Type İI Interaction Plot for the Data in Table 1.1 Using Humidity as a Baseline. Lines Are Those Estimated by Tukey's Model.**

Before discussing Mandel's test for interaction, the two possible Type II interaction plots of the cell means in Table 1.7 are first examined. These plots are given in Figures 1.7 and 1.8.

    Examination of the Type II plot in Figure 1.7 reveals no information about a possible model for the experiment, whereas examination of the Type II plot in Figure 1.8 suggests that the response variable $\mu_{ij}$ is a linear function of $\beta_j$ for each value of $i$ with a different slope parameter for each line. This is why Mandel's model is called a "bundle-of-straight-lines" model.

    For data which can be modeled by Tukey's model, it does not matter whether the $\mu_{ij}$'s are plotted against $\tau_i$ or plotted against $\beta_j$, the resulting plots will consist of straight lines which intersect in a common point; but for Mandel's model it makes a difference. Had the plot in Figure 1.7 yielded a bundle of straight lines, an appropriate model would be of the form,

$$\mu_{ij} = \mu + \tau_i + \beta_j + \tau_i \gamma_j \qquad (1.8.3)$$

**1.8.1
Interaction
Plots for Cell
Means
Satisfying
Mandel's
Model**

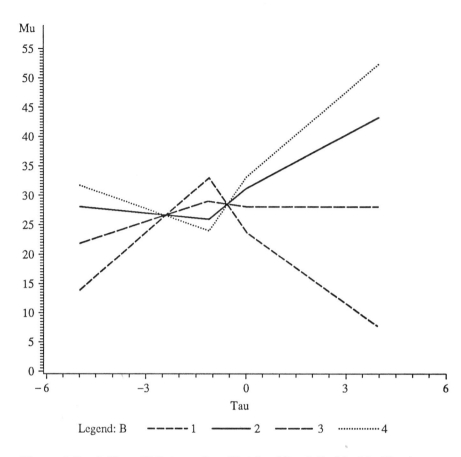

**Figure 1.7   A Type II Interaction Plot for Mandel's Model, Plotting $\mu_{ij}$ Against $\tau_i$.**

or equivalently,

$$\mu_{ij} = \mu + \beta_j + \tau_i \eta_j$$

where $\eta_j = 1 + \gamma_j$.

**1.8.2
Mandel's Test
for Interaction**

Next Mandel's test of the no-interaction hypothesis, namely $H_0$: $\alpha_1 = \alpha_2 = \cdots = \alpha_t$ in model (1.8.1), is examined. The sum of squares for testing $H_0$ is given by

$$\text{SS}_{H_0} = \left( \sum_{i=1}^{t} \hat{\alpha}_i^2 \right)\left( \sum_{j=1}^{b} \hat{\beta}_j^2 \right) \tag{1.8.4}$$

where

$$\hat{\beta}_j = \bar{y}_{\bullet j} - \bar{y}_{\bullet \bullet} \qquad j = 1, 2, \ldots, b$$

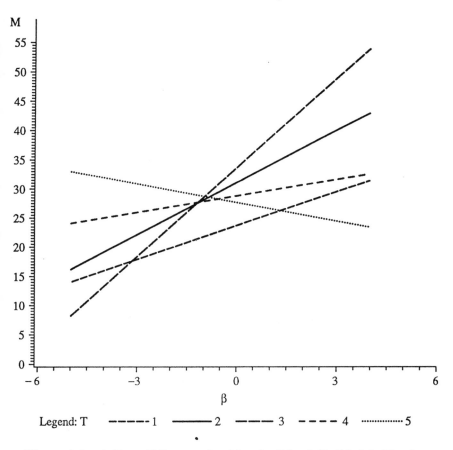

**Figure 1.8** **A Type II Interaction Plot for Mandel's Model, Plotting** $\mu_{ij}$ **Against** $\beta_j$.

and

$$\hat{\alpha}_i = \sum_{j=1}^{b} (y_{ij} - \bar{y}_{\cdot j})\hat{\beta}_j \left/ \sum_{j=1}^{b} \hat{\beta}_j^2 \right. \qquad (1.8.5)$$

The residual sum of squares after fitting Mandel's model (1.8.1) is equal to

$$SS_R = \sum_{ij} z_{ij}^2 - SS_{H_0} \qquad (1.8.6)$$

where

$$z_{ij} = y_{ij} - \bar{y}_{i\cdot} - \bar{y}_{\cdot j} + \bar{y}_{\cdot\cdot}$$

When $H_0$ is true, $SS_{H_0}/\sigma^2$ has a sampling distribution which is chi-square with $t - 1$ degrees of freedom. Also, when $H_0$ is true, $SS_{H_0}$ is distributed independently of $SS_R$, and the sampling distribution of

$SS_R/\sigma^2$ is chi-square with $(b-1)(t-1) - (t-1) = (t-1)(b-2)$ degrees of freedom. Thus, when $H_0$ is true in model (1.8.1)

$$F = \frac{SS_{H_0}/(t-1)}{SS_R/(t-1)(b-2)}$$

has a sampling distribution, which is $F_{t-1,(t-1)(b-2)}$. The hypothesis $H_0: \alpha_1 = \alpha_2 = \cdots = \alpha_t$ is rejected if $F > F_{\alpha,t-1,(t-1)(b-2)}$.

If Mandel's model is of the form given in (1.8.3), then the sum of squares for testing $H_0: \gamma_1 = \gamma_2 = \cdots = \gamma_b$ is

$$SS_{H_0} = \left(\sum_{i=1}^{t} \hat{\tau}_i^2\right)\left(\sum_{j=1}^{b} \hat{\gamma}_j^2\right) \tag{1.8.7}$$

where

$$\hat{\tau}_i = \bar{y}_{i\cdot} - \bar{y}_{\cdot\cdot}$$

and

$$\hat{\gamma}_j = \sum_{i=1}^{t} (y_{ij} - \bar{y}_{i\cdot})\hat{\tau}_i \Big/ \sum \hat{\tau}_i^2 \tag{1.8.8}$$

The residual sum of squares after fitting Mandel's model (1.8.3) is

$$SS_R = \sum_{ij} z_{ij}^2 - \left(\sum_i \hat{\tau}_i^2\right)\left(\sum_j \hat{\gamma}_j^2\right) \tag{1.8.9}$$

and has $(t-2)(b-1)$ degrees of freedom.

For model (1.8.3), the hypothesis $H_0: \gamma_1 = \gamma_2 = \cdots = \gamma_b$ is rejected if

$$F = \frac{\left(\sum \hat{\tau}_i^2\right)\left(\sum \hat{\gamma}_j^2\right)/(b-1)}{\left[\sum_{ij} z_{ij}^2 - \left(\sum \hat{\tau}_i^2\right)\left(\sum \hat{\gamma}_j^2\right)\right]/(t-2)(b-1)} > F_{\alpha,b-1,(t-2)(b-1)}$$

**1.8.3
Computing
Mandel's
Test Using
Statistical
Software**

Mandel's analysis for model (1.8.1) can also be obtained from many statistical computing packages by following the steps below:

Step 1. Fit the additive model $y_{ij} = \mu + \tau_i + \beta_j + \epsilon_{ij}$ as in Section 1.6.1 and take the solution to the normal equations satisfying $\sum \hat{\tau}_i = \sum \hat{\beta}_j = 0$ and compute the residuals

$$z_{ij} = y_{ij} - \bar{y}_{i\cdot} - \bar{y}_{\cdot j} + \bar{y}_{\cdot\cdot}$$

Step 2. Fit each of the models

$$z_{ij} = \alpha_i \hat{\beta}_j + e_{ij} \qquad i = 1, 2, \ldots, t \qquad (1.8.10)$$

Each of the $t$ models in (1.8.10) is a simple linear regression model with an intercept at zero and hence $\hat{\alpha}_i = \sum_j \hat{\beta}_j z_{ij} / \sum_j \hat{\beta}_j^2$, which is algebraically equivalent to $\hat{\alpha}_i = \sum_j \hat{\beta}_j (y_{ij} - \bar{y}_{\cdot j}) / \sum \hat{\beta}_j^2$. The regression sum of squares from the $i$th model is

$$\mathrm{SS}_{\mathrm{REG}_i} = \left( \sum_j \hat{\beta}_j z_{ij} \right)^2 \bigg/ \sum_j \hat{\beta}_j^2 = \hat{\alpha}_i^2 \left( \sum_j \hat{\beta}_j^2 \right)$$

$$i = 1, 2, \ldots, t$$

The residual sum of squares for the $i$th model is

$$\mathrm{SS}_{R_i} = \sum_j z_{ij}^2 - \mathrm{SS}_{\mathrm{REG}_i} = \sum_j z_{ij}^2 - \hat{\alpha}_i^2 \sum_j \hat{\beta}_j^2$$

The sum of squares due to $H_0$: $\alpha_1 = \alpha_2 = \cdots = \alpha_t$ is obtained by pooling the $\mathrm{SS}_{\mathrm{REG}_i}$ over all $i$ and is given by

$$\mathrm{SS}_{H_0} = \sum_i \mathrm{SS}_{\mathrm{REG}_i} = \sum_i \hat{\alpha}_i^2 \sum_j \hat{\beta}_j^2$$

The residual sum of squares from all models is obtained by pooling the $\mathrm{SS}_{R_i}$ over all $i$ and is given by

$$\mathrm{SS}_R = \sum_i \mathrm{SS}_{R_i} = \sum_{ij} z_{ij}^2 - \sum_i \hat{\alpha}_i^2 \sum_j \hat{\beta}_j^2$$

Step 3. Calculate $F$ by

$$F = \frac{\mathrm{SS}_{H_0}/(t-1)}{\mathrm{SS}_R/(b-2)(t-1)}$$

and reject $H_0$, if

$$F > F_{\alpha, t-1, (b-2)(t-1)}$$

Mandel's analysis can be obtained with SAS® by following these steps:
Step 1. Fit model (1.6.5) by using

PROC GLM; CLASSES T B;

MODEL Y = T B/SOLUTION;

**1.8.4
Computing
Mandel's Test
Using SAS®**

Step 2. Find the solution vector satisfying

$$\sum \hat{\tau}_i = \sum \hat{\beta}_j = 0$$

by using equations (1.6.6).

Step 3. Make a new set of data cards by adding the new solutions, $\hat{\tau}_i^*$ and $\hat{\beta}_j^*$ to the data card which contains the treatment combination $T_i$ and $B_j$, and fit the $t$ models,

$$y_{ij} = \mu + \tau_i + \beta_j + \lambda\alpha_i\beta_j + \epsilon_{ij} \qquad i = 1, 2, \ldots, t$$

simultaneously, by using

PROC GLM; CLASSES T B;

MODEL Y = T B T*BHAT/SOLUTION;

The $F$-test in either the Type I or Type III analysis that corresponds to the row labeled by T*BHAT is Mandel's test for nonadditivity.

**1.8.5**
**Examples**
**Illustrating**
**Mandel's**
**Model**

Consider once again the data in Table 1.1. To begin, the data must first be analyzed by using SAS® GLM and the commands:

PROC GLM;

CLASSES TEMP HUMIDITY;

MODEL HEIGHT = TEMP HUMIDITY/SOLUTION;

As in Section 1.6.3, the results from this run are used to obtain the data set in Table 1.3. The data in Table 1.1 will be analyzed by using two Mandel-type models. In Section 1.8.5.1 different lines are obtained for each humidity level, while in Section 1.8.5.2 different lines are obtained for each temperature level.

**1.8.5.1 Mandel's Bundle-of-Straight-Lines Model for Each Level of Humidity**   For this case the data in Table 1.3 are analyzed with SAS® GLM by using the commands:

PROC GLM;

CLASSES TEMP HUMIDITY;

MODEL HEIGHT = TEMP HUMIDITY

THAT*HUMIDITY/SOLUTION;

This fits a different straight-line model for each humidity level.

The Type I and Type III analysis of variance tables generated by the above commands are given in Table 1.9. The results from the SOLUTION option are given in Table 1.10.

**Table 1.9  Type I and Type III Analysis of Variance Tables for Mandel's Model with Temperature as a Baseline**

| Source | DF | Sum of squares | | Mean square |
|---|---|---|---|---|
| Model | 10 | 2501.14963514 | | 250.11496351 |
| Error | 9 | 110.21236486 | | 12.24581832 |
| Corrected Total | 19 | 2611.36200000 | | |
| Model F = | 20.42 | | | PR > F = 0.0001 |

| R-square | C.V. | Root MSE | | Y mean |
|---|---|---|---|---|
| 0.957795 | 13.9808 | 3.49940257 | | 25.03000000 |

| Source | DF | Type I SS | F value | PR > F |
|---|---|---|---|---|
| Temp | 4 | 136.61700000 | 2.79 | 0.0927 |
| Humidity | 3 | 2074.29800000 | 56.46 | 0.0001 |
| THAT*Humidity | 3 | 290.23463514 | 7.90 | 0.0069 |

| Source | DF | Type III SS | F value | PR > F |
|---|---|---|---|---|
| Temp | 3 | 0.00000000 | 0.00 | 1.0000 |
| Humidity | 3 | 2074.29800000 | 56.46 | 0.0001 |
| THAT*Humidity | 3 | 290.23463514 | 7.90 | 0.0069 |

The $F$-value ($F = 7.90$) corresponding to THAT*HUMIDITY in Table 1.9 is Mandel's test for nonadditivity, and $H_0 : \gamma_1 = \gamma_2 = \cdots = \gamma_5$ in (1.8.3) is rejected at the .0069 significance level. Also, from Table 1.9 the value of $SS_{H_0}$ in (1.8.7) is $SS_{H_0} = 290.235$, and the value of $SS_R$ in (1.8.9) is $SS_R = 110.212$ and is based on 9 degrees of freedom.

**Table 1.10  Results from SOLUTION Option of SAS® GLM Using Mandel's Model with Temperature as the Baseline**

| Parameter | | Estimate | T for $H_0$: Parameter $= 0$ | PR > \|T\| | STD Error of Estimate |
|---|---|---|---|---|---|
| Intercept | | 49.27636275 B | 17.42 | 0.0001 | 2.82803323 |
| Temp | 50 | − 19.75070123 B | − 4.76 | 0.0010 | 4.15013431 |
| | 60 | − 18.90534048 B | − 4.68 | 0.0011 | 4.03654506 |
| | 70 | − 10.52858392 B | − 3.46 | 0.0072 | 3.04587794 |
| | 80 | − 2.99718812 B | − 1.19 | 0.2657 | 2.52557720 |
| | 90 | 0.00000000 B | . | . | . |
| Humidity | 20 | − 26.28000000 B | − 11.87 | 0.0001 | 2.21321651 |
| | 40 | − 20.30000000 B | − 9.17 | 0.0001 | 2.21321651 |
| | 60 | − 8.66000000 B | − 3.91 | 0.0035 | 2.21321651 |
| | 80 | 0.00000000 B | . | . | . |
| THAT*Humidity | 20 | − 3.74274065 B | − 4.42 | 0.0017 | 0.84681107 |
| | 40 | − 3.09200173 B | − 3.65 | 0.0053 | 0.84681107 |
| | 60 | − 1.46141403 B | − 1.73 | 0.1185 | 0.84681107 |
| | 80 | 0.00000000 B | . | . | . |

From the last four rows of Table 1.10, one can find the values of the $\hat{\gamma}_j$'s defined in (1.8.8). The estimates given in Table 1.10 satisfy $\hat{\gamma}_b = 0$, while those in (1.8.8) satisfy $\sum \hat{\gamma}_j = 0$. Thus the $\hat{\gamma}_j$'s satisfying $\sum \hat{\gamma}_j = 0$ are

$$\hat{\gamma}_1 = -3.743 - a$$

$$\hat{\gamma}_2 = -3.092 - a$$

$$\hat{\gamma}_3 = -1.461 - a$$

$$\hat{\gamma}_4 = .000 - a$$

where

$$a = [(-3.743) + (-3.092) + (-1.461) + .000]/4$$
$$= -2.074$$

Hence

$$\hat{\gamma}_1 = -1.669$$

$$\hat{\gamma}_2 = -1.018$$

$$\hat{\gamma}_3 = .613$$

$$\hat{\gamma}_4 = 2.074$$

The ERROR MEAN SQUARE = 12.246 for this Mandel's model applied to these data. Recall that for Tukey's model, the ERROR MEAN SQUARE = 10.163. Since Mandel's model has a larger error mean square than Tukey's model, one could hardly recommend this Mandel's model over Tukey's model. In Section 1.9 a statistical test for determining if Mandel's model provides a significantly better fit to the data than that provided by Tukey's model is discussed.

**1.8.5.2 Mandel's Bundle-of-Straight-Lines Model for Each Level of Temperature**   Next consider using Mandel's model so that a different straight line is obtained for each temperature level. This can be done by analyzing the data in Table 1.3 by using the SAS® commands:

PROC GLM;

CLASSES TEMP HUMIDITY;

MODEL HEIGHT = TEMP HUMIDITY TEMP*HHAT/SOLUTION;

The Type I and Type III analysis of variance tables are given in Table 1.11. The results from the SOLUTION option are in Table 1.12.

From Table 1.11, Mandel's test for nonadditivity is $F = 17.99$ and $H_0$: $\alpha_1 = \alpha_2 = \alpha_3 = \alpha_4 = \alpha_5$ in (1.8.1) can be rejected at the .0005 significance level. The value of $SS_{H_0}$ in (1.8.4) is $SS_{H_0} = 360.383$ and has 4 degrees of freedom. The value of $SS_R$ in (1.8.6) is $SS_R = 40.064$ with 8 degrees of freedom. The ERROR MEAN SQUARE for this model is 5.008 and is based on 8 degrees of freedom, which is about half

**Table 1.11 Type I and Type III Analysis of Variance Tables for Mandel's Model with Humidity as a Baseline**

| Source | DF | Sum of squares | | Mean square |
|---|---|---|---|---|
| Model | 11 | 2571.29824885 | | 233.75438626 |
| Error | 8 | 40.06375115 | | 5.00796889 |
| Corrected Total | 19 | 2611.36200000 | | |
| Model F = | 46.68 | | | PR > F = 0.0001 |
| R-square | C.V. | Root MSE | | Y mean |
| 0.984658 | 8.9407 | 2.23784917 | | 25.03000000 |
| Source | DF | Type I SS | F value | PR > F |
| Temp | 4 | 136.61700000 | 6.82 | 0.0108 |
| Humidity | 3 | 2074.29800000 | 138.07 | 0.0001 |
| HHAT*Temp | 4 | 360.38324885 | 17.99 | 0.0005 |
| Source | DF | Type III SS | F value | PR > F |
| Temp | 4 | 136.61700000 | 6.82 | 0.0108 |
| Humidity | 2 | 0.00000000 | 0.00 | 1.0000 |
| HHAT*Temp | 4 | 360.38324885 | 17.99 | 0.0005 |

of the ERROR MEAN SQUARE that was given by Tukey's model. A comparison of these two error mean squares would seem to indicate that Mandel's model using humidity as a baseline provides a better fit to the data than Tukey's model. There is more discussion on such comparisons in Section 1.9.

From the last five lines of Table 1.12, the values of the $\hat{\alpha}_i$'s in (1.8.5) which satisfy $\sum \hat{\alpha}_i = 0$ can be determined. One gets

$$\hat{\alpha}_1 = -1.0426 - b$$

$$\hat{\alpha}_2 = -.9897 - b$$

$$\hat{\alpha}_3 = -1.0236 - b$$

$$\hat{\alpha}_4 = -.4304 - b$$

$$\hat{\alpha}_5 = .000 - b$$

where

$$b = [(-1.0426) + (-.9897) + (-1.0236) + (-.4304) + (.000)]/5$$
$$= -.6973$$

Thus

$$\hat{\alpha}_1 = -.3453$$

$$\hat{\alpha}_2 = -.2924$$

$$\hat{\alpha}_3 = -.3263$$

**Table 1.12 Results from SOLUTION Option of SAS® GLM Using Mandel's Model with Humidity as the Baseline**

| Parameter | | Estimate | T for $H_0$: Parameter $=0$ | PR $> |T|$ | STD Error of Estimate |
|---|---|---|---|---|---|
| Intercept | | 51.86417376 B | 26.45 | 0.0001 | 1.96085896 |
| Temp | 50 | $-6.42500000$ B | $-4.06$ | 0.0036 | 1.58239832 |
| | 60 | $-6.15000000$ B | $-3.89$ | 0.0046 | 1.58239832 |
| | 70 | $-3.42500000$ B | $-2.16$ | 0.0624 | 1.58239832 |
| | 80 | $-0.97500000$ B | $-0.62$ | 0.5549 | 1.58239832 |
| | 90 | 0.00000000 B | . | . | . |
| Humidity | 20 | $-44.60401784$ B | $-15.15$ | 0.0001 | 2.94496333 |
| | 40 | $-34.45439734$ B | $-14.09$ | 0.0001 | 2.44597959 |
| | 60 | $-14.69827985$ B | $-8.90$ | 0.0001 | 1.65149441 |
| | 80 | 0.00000000 B | . | . | . |
| HHAT*Temp | 50 | $-1.04259610$ B | $-6.71$ | 0.0002 | 0.15538004 |
| | 60 | $-0.98968422$ B | $-6.37$ | 0.0002 | 0.15538004 |
| | 70 | $-1.02364029$ B | $-6.59$ | 0.0002 | 0.15538004 |
| | 80 | $-0.43038416$ B | $-2.77$ | 0.0243 | 0.15538004 |
| | 90 | 0.00000000 B | . | . | . |

$$\hat{\alpha}_4 = .2669$$

$$\hat{\alpha}_5 = .6973$$

Figure 1.9 gives a Type II interaction plot of the data in Table 1.1 along with the estimated regression lines given by Mandel's model with a line for each temperature. To illustrate the plot of the line corresponding to Temperature $= 80$ ($i = 4$), the estimated mean for Humidity $= 20$ ($j = 1$) is

$$\hat{\mu}_{41} = \hat{\mu} + \hat{\tau}_4 + \hat{\beta}_1 + \hat{\alpha}_4\hat{\beta}_1$$

$$= 25.03 + 2.420 + (-12.47) + (.2669)(-12.47)$$

$$= 11.65$$

and in a similar manner one can find that the estimated mean for Humidity $= 80$ ($j = 4$) is

$$\hat{\mu}_{44} = \hat{\mu} + \hat{\tau}_4 + \hat{\beta}_4 + \hat{\alpha}_4\hat{\beta}_4$$

$$= 25.03 + 2.42 + 13.81 + (.2669)(13.81)$$

$$= 44.95$$

Examination of the estimated regression lines in Figure 1.9 would seem to indicate that there is very little interaction between Temperatures 50, 60, and 70 and all Humidity levels. This is because these three lines are nearly parallel to each other. It is interesting to note that if one computes Mandel's test for interaction on the data in Table 1.1 with Temperature levels 80 and 90 deleted, one gets $F = .06$ with 2 and 4 degrees of freedom, which is not significant.

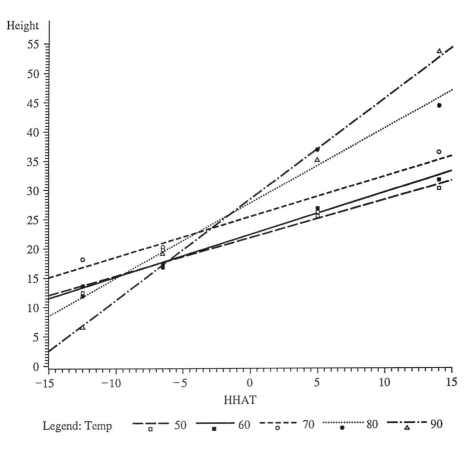

**Figure 1.9    A Type II Interaction Plot for the Data in Table 1.1 Using Humidity as a Baseline. Lines Are Those Predicted by Mandel's Model.**

It is clear that Tukey's model is a special case of Mandel's model. That is, data that can be modeled by Tukey's model give a bundle of straight lines with the additional restriction that all those lines intersect in a single point. In Mandel's model (1.8.1) define $\theta_i$ by

$$\alpha_i = \lambda\tau_i + \theta_i \qquad i = 1, 2, \ldots, t$$

then Mandel's model can be written as

$$\mu_{ij} = \mu + \tau_i + \beta_j + \lambda\tau_i\beta_j + \theta_i\beta_j$$
$$i = 1, 2, \ldots, t, \quad j = 1, 2, \ldots, b \qquad (1.9.1)$$

**1.9 COMPARING MANDEL'S MODEL AND TUKEY'S MODEL**

where the first four terms on the right side of this model consist of Tukey's model.

Thus, if $\theta_1 = \theta_2 = \cdots = \theta_t$ and $\lambda \neq 0$, Mandel's model simplifies to Tukey's model. To estimate the parameters in the above model one can estimate $\lambda$ as before; i.e.,

$$\hat{\lambda} = \sum_{ij} \hat{\tau}_i \hat{\beta}_j z_{ij} \bigg/ \left( \sum_i \hat{\tau}_i^2 \sum_j \hat{\beta}_j^2 \right)$$

and take

$$\hat{\theta}_i = \hat{\alpha}_i - \hat{\lambda}\hat{\tau}_i$$

where $\hat{\alpha}_i = \sum_j \hat{\beta}_j z_{ij} / \sum \hat{\beta}_j^2$ for each $i$.

A decision as to whether Mandel's model provides a significantly better fit to the data than Tukey's model can be made by testing $H_0$: $\theta_1 = \theta_2 = \cdots = \theta_t$ in (1.9.1). In this regard, note that the regression sum of squares (1.8.4) for testing $\alpha_1 = \alpha_2 = \cdots = \alpha_t$ can be partitioned into two parts.

The first part is $SSN = (\sum_{ij} \hat{\tau}_i^2 \hat{\beta}_j z_{ij})^2 / (\sum \hat{\tau}_i \sum \hat{\beta}_j^2)$ which has one degree of freedom for testing $\lambda = 0$, and the second part is $SSC = \sum_i \hat{\alpha}_i^2 \cdot \sum_j \hat{\beta}_j^2 - SSN$ with $t-2$ degrees of freedom for testing $H_0$: $\theta_1 = \theta_2 = \cdots = \theta_t$. Such a test has been called a test for concurrency. If one rejects (1) $\lambda = 0$ and accepts (2) $H_0$: $\theta_1 = \theta_2 = \cdots = \theta_t$, the bundle of straight lines is called concurrent; i.e., they intersect in a common point. If one rejects both (1) and (2), the bundle of straight lines is called nonconcurrent.

To use SAS® GLM for an analysis that tests the concurrency hypothesis described above, one uses the following commands on the data that have been augmented by the values of the estimates in (1.6.6).

PROC GLM; CLASSES T B;

MODEL Y = T B THAT*BHAT T*BHAT/SOLUTION;

For example, the data in Table 1.1 were analyzed in this manner with SAS® GLM using the following commands on the data in Table 1.3.

PROC GLM;

CLASSES TEMP HUMIDITY;

MODEL HEIGHT = TEMP HUMIDITY

THAT*HHAT TEMP*HHAT/SOLUTION;

The Type I and Type III analysis of variance tables from this analysis are given in Table 1.13.

In Table 1.13 the $F = 4.77$ in the row labeled HHAT*TEMP is the test for concurrency. This value of $F$ is significant at the .0343 level, indicating that Mandel's model, which allows a different line for each temperature level, provides a significantly better fit to the data in Table 1.1 than does Tukey's model.

**Table 1.13   Type I and Type III Analysis of Variance Tables
from SAS® GLM. A Test of Concurrency**

| Source | DF | Sum of squares | | Mean square |
|---|---|---|---|---|
| Model | 11 | 2571.29824885 | | 233.75438626 |
| Error | 8 | 40.06375115 | | 5.00796889 |
| Corrected Total | 19 | 2611.36200000 | | |
| Model F = | 46.68 | | | PR > F = 0.0001 |
| R-square | C.V. | Root MSE | | Y mean |
| 0.984658 | 8.9407 | 2.23784917 | | 25.03000000 |
| Source | DF | Type I SS | F value | PR > F |
| Temp | 4 | 136.61700000 | 6.82 | 0.0108 |
| Humidity | 3 | 2074.29800000 | 138.07 | 0.0001 |
| THAT*HHAT | 1 | 288.65200865 | 57.64 | 0.0001 |
| HHAT*Temp | 3 | 71.73124020 | 4.77 | 0.0343 |
| Source | DF | Type III SS | F value | PR > F |
| Temp | 4 | 136.61700000 | 6.82 | 0.0108 |
| Humidity | 2 | 0.00000000 | 0.00 | 1.0000 |
| THAT*HHAT | 0 | 0.00000000 | . | . |
| HHAT*Temp | 3 | 71.73124020 | 4.77 | 0.0343 |

There have been several extensions of these types of tests to other interaction models by using different functions to model the interaction terms and also to other designs such as the $n$-way cross-classified treatment structures and Latin square treatment structures. Milliken and Graybill (1970) introduced a generalized procedure which allows one to use any known function of the main-effect parameters to model the interaction. Their generalization is illustrated with an example.

**1.10
GENERALIZED
INTERACTION
MODELS**

Suppose the cell means model is given by

$$\mu_{ij} = \mu + \tau_i + \beta_j + \gamma_{ij}$$

and the experimenter believes the interaction, if it exists, can be modeled by $\lambda\tau_i e^{-\beta_j}$. A test for interaction of this form is obtained by testing $\lambda = 0$ in the model

$$\mu_{ij} = \mu + \tau_i + \beta_j + \lambda\tau_i e^{-\beta_j} \qquad (1.10.1)$$

The test statistic is obtained by first fitting the additive model,

$$y_{ij} = \mu + \tau_i + \beta_j + \epsilon_{ij}$$

and then fitting the model

$$z_{ij} = \lambda\hat{\tau}_i e^{-\hat{\beta}_j} + e_{ij}$$

The estimate of $\lambda$ is

$$\hat{\lambda} = \sum_{ij}\left(\hat{\tau}_i e^{-\hat{\beta}_j}z_{ij}\right)\Bigg/ \sum_{ij}\left(\hat{\tau}_i e^{-\hat{\beta}_j}\right)^2$$

The sum of squares for nonadditivity is

$$\text{NSS} = \left[\sum_{ij}\hat{\tau}_i e^{-\hat{\beta}_j}z_{ij}\right]^2\Bigg/ \sum_{ij}\left(\hat{\tau}_i e^{-\hat{\beta}_j}\right)^2$$

and one rejects $H_0$: $\lambda = 0$, if

$$F = \frac{\text{NSS}/1}{(\sum_{ij} z_{ij}^2 - \text{NSS})/[(b-1)(t-1)-1]} > F_{\alpha,1,(b-1)(t-1)-1}$$

To obtain this analysis using existing statistical software, one would need to construct a new variable, call it $x_{ij}$, such that $x_{ij} = \hat{\tau}_i e^{-\hat{\beta}_j}$ for all $i$ and $j$, and augment the original data set with the values of this new variable, after which one can proceed as in Section 1.6.

To illustrate, the SAS® statements that would be required to fit model (1.10.1) to the data in Table 1.3 are

PROC GLM;

CLASSES TEMP HUMIDITY;

MODEL HEIGHT = TEMP HUMIDITY X;

where X = THAT*EXP(-HHAT).

**1.11 CHARACTER-ISTIC ROOT TEST FOR INTERACTION**

The procedures discussed in the preceding sections have two things in common:

1. The tests are based on models where the interaction is a known function of the two sets of treatment effects, and

2. The main problem was to determine whether interaction is present.

If one such test indicates interaction is present, many statisticians have recommended that the data be transformed, using a nonlinear transformation, so that the resulting data can be described by an additive model. For example, if Tukey's model has been determined to model the data adequately, then it is often the case that the logarithms of the $y_{ij}$'s can be modeled by an additive model. We do not usually recommend that one transform data just because one wants to model the data with an additive model. Our reasons for not recommending transformations are

1. If one believed the error term was additive in the model used for the first test for interaction, then it cannot still be additive after transforming the data by a nonlinear transformation.

2. The data are usually much easier to interpret in terms of the original units than they are in terms of the transformed units.

3. It is not necessary to transform the data. If the data can be modeled, then interpretations can come from the selected model.

In this section a model is introduced that is often quite "good" for modeling two-way data with interaction, and the model does not require the experimenter to model the interaction in terms of main-effect parameters as was required for the tests in previous sections. The model is called the *multiplicative interaction model*. The multiplicative interaction model is defined by

**1.11.1 A Multiplicative Interaction Model**

$$\mu_{ij} = \mu + \tau_i + \beta_j + \lambda \alpha_i \gamma_j \qquad (1.11.1)$$

where it is assumed that

$$\sum_i \tau_i = \sum_j \beta_j = \sum_i \alpha_i = \sum_j \gamma_j = 0 \qquad (1.11.2)$$

and

$$\sum_i \alpha_i^2 = \sum_j \gamma_j^2 = 1$$

These assumptions can be made without any loss of generality in the model; the assumptions are made only to limit the number of possible values that the parameters may have when satisfying model (1.11.1).

In model (1.11.1), note that if $\alpha_i = \alpha_{i'}$ for some $i \neq i'$, then there is no interaction between row treatments $i$ and $i'$ and the column treatments. This is true because for any $j$ and $j'$,

$$
\begin{aligned}
\mu_{ij} - \mu_{i'j} - \mu_{ij'} + \mu_{i'j'} &= (\mu + \tau_i + \beta_j + \lambda \alpha_i \gamma_j) \\
&\quad - (\mu + \tau_{i'} + \beta_j + \lambda \alpha_{i'} \gamma_j) \\
&\quad - (\mu + \tau_i + \beta_{j'} + \lambda \alpha_i \gamma_{j'}) \\
&\quad + (\mu + \tau_{i'} + \beta_{j'} + \lambda \alpha_{i'} \gamma_{j'}) \\
&= \lambda \alpha_i \gamma_j - \lambda \alpha_{i'} \gamma_j - \lambda \alpha_i \gamma_{j'} + \lambda \alpha_{i'} \gamma_{j'} \\
&= \lambda (\alpha_i - \alpha_{i'}) \gamma_j - \lambda (\alpha_i - \alpha_{i'}) \gamma_{j'} \\
&= \lambda (\alpha_i - \alpha_{i'}) (\gamma_j - \gamma_{j'}) \\
&= 0 \quad \text{if } \alpha_i = \alpha_{i'}
\end{aligned}
$$

Similarly, if $\gamma_j = \gamma_{j'}$ for some $j \neq j'$, then there is no interaction between column treatments $j$ and $j'$ and the row treatments.

There are many situations where the multiplicative interaction model is more appropriate than either Tukey's model or Mandel's model, even though all three models have similar forms. Both Tukey's and Mandel's models are special cases of the multiplicative interaction model. Some situations where the multiplicative interaction model is more appropriate than Tukey's or Mandel's models are given below:

1. When interaction is present in the model, but there are no main-effect differences in either the $T$ treatments or the $B$ treatments.

2. When interaction is present in only one cell. This might be the case if the observation in a given cell is an outlier. It might also be the case when one particular treatment combination gives strange results when applied to the experimental unit. This has often been seen to be the case when both sets of treatments have a control level. The combination of both controls together often gives a result which does not allow all of the data to be modeled by an additive model, i.e., this one cell is responsible for all of the interaction in the data.

3. When all of the interaction is present in only one row or one column. This is often the case when one of the treatment levels is a control. The control may act differently than all other levels of that factor, in that the other levels do not interact with the levels of the remaining factor. This situation may also be the case when there are multiple outliers in the data which occur in the same row or the same column.

**1.11.2
Examples**

To illustrate the applicability of the multiplicative interaction model and how it can be used to help interpret the results of an experiment, several contrived examples are examined.

Consider once again the true cell means of an additive model given in Table 1.6. For these cell mean parameters, possible values of the parameters in model (1.11.1) are

$$\mu = 29, \qquad \tau' = [-5 \quad 2 \quad 4 \quad 0 \quad -1]$$
$$\beta' = [-5 \quad 2 \quad -1 \quad 4], \qquad \lambda = 0$$

The $\alpha$'s and $\gamma$'s could have any values satisfying the restrictions in (1.11.2).

Next consider the cell means in Table 1.14, which are additive except for one cell. Before reading further, examine Table 1.14 and try to determine which treatment combination is responsible for the interaction in this table.

One set of values of the parameters in model (1.11.1) satisfying the restrictions in (1.11.2) which describe the cell means in Table 1.14 is

$$\mu = 28.55$$

$$\tau' = [-4.55 \quad .20 \quad 4.45 \quad .45 \quad -.55]$$

**Table 1.14   Cell Means with One Cell Containing the Interaction**

|       | $B_1$ | $B_2$ | $B_3$ | $B_4$ |
|-------|-------|-------|-------|-------|
| $T_1$ | 19    | 26    | 23    | 28    |
| $T_2$ | 26    | 24    | 30    | 35    |
| $T_3$ | 28    | 35    | 32    | 37    |
| $T_4$ | 24    | 31    | 28    | 33    |
| $T_5$ | 23    | 30    | 27    | 32    |

$$\beta' = [-4.55 \quad .65 \quad -.55 \quad 4.45]$$

$$\lambda = 6.9714$$

$$\alpha' = [-.2236 \quad .8944 \quad -.2236 \quad -.2236 \quad -.2236]$$

$$\gamma' = [.2887 \quad -.8660 \quad .2887 \quad .2887]$$

Note that $\alpha_1 = \alpha_3 = \alpha_4 = \alpha_5$, which implies there is no interaction between row treatments 1, 3, 4, and 5 and the column treatments. Also note that $\gamma_1 = \gamma_3 = \gamma_4$, which implies there is no interaction between column treatments 1, 3, and 4 and the row treatments. Thus, since $\lambda \neq 0$, one is able to determine there is interaction in the table, and the values of $\alpha$ and $\gamma$ given above imply that the interaction exists only because of the mean in the (2,2) cell. That is, the combination of $T_2$ and $B_2$ gives a response different from what would be expected if the two sets of treatment combinations were additive. Indeed, if the value in the (2,2) cell were 33, there would be no interaction in these cell means.

Next consider the cell means given in Table 1.15.

One set of parameters for model (1.11.1), which describes the cell means in Table 1.15, is

$$\mu = 29$$

$$\tau' = [-5 \quad 2 \quad 4 \quad 0 \quad -1]$$

$$\beta' = [-3.2 \quad .6 \quad -1.0 \quad 3.6]$$

$$\lambda = 10.5537$$

$$\alpha' = [.8944 \quad -.2236 \quad -.2236 \quad -.2236 \quad -.2236]$$

$$\gamma' = [.7775 \quad -.6047 \quad .000 \quad -.1728]$$

Since $\alpha_2 = \alpha_3 = \alpha_4 = \alpha_5$, there is no interaction between row treatments 2, 3, 4, and 5 and the column treatments. Thus the cells in the first row give responses different from those which would be expected with an additive model.

The cell means given in Table 1.7 were generated by assuming Tukey's model. Values of the parameters in model (1.11.1), which fit

**Table 1.15   Cell Means with One Row Causing the Interaction**

| $\mu_{ij}$ | $B_1$ | $B_2$ | $B_3$ | $B_4$ |
|---|---|---|---|---|
| $T_1$ | 28 | 19 | 23 | 26 |
| $T_2$ | 26 | 33 | 30 | 35 |
| $T_3$ | 28 | 35 | 32 | 37 |
| $T_4$ | 24 | 31 | 28 | 33 |
| $T_5$ | 23 | 30 | 27 | 32 |

the values in Table 1.7, are

$$\mu = 29$$

$$\tau' = [-5 \quad 2 \quad 4 \quad 0 \quad -1]$$

$$\beta' = [-5 \quad 2 \quad -1 \quad 4]$$

$$\lambda = 46.0$$

$$\alpha' = [.7372 \quad -.2949 \quad -.5898 \quad 0 \quad .1474]$$

$$\gamma' = [.7372 \quad -.2949 \quad .1474 \quad -.5898]$$

Note that none of the $\alpha$'s are the same and none of the $\gamma$'s are the same; this implies that all row treatments and column treatments interact in these data. Note that the correlation between the elements in the $\tau$ vector and the corresponding elements in the $\alpha$ vector is

$$\rho_{\alpha, \tau} = \sum_i (\alpha_i - \bar{\alpha}_{\bullet})(\tau_i - \bar{\tau}_{\bullet}) \left/ \left[ \sum (\alpha_i - \bar{\alpha}_{\bullet})^2 \sum (\tau_i - \bar{\tau}_{\bullet})^2 \right]^{1/2} \right. = -1$$

and the correlation between the elements in the $\beta$ vector and those in the $\gamma$ vector is

$$\rho_{\beta, \gamma} = \sum_j (\beta_j - \bar{\beta}_{\bullet})(\gamma_j - \bar{\gamma}_{\bullet}) \left/ \left[ \sum (\beta_j - \bar{\beta}_{\bullet})^2 \sum (\gamma_j - \bar{\gamma}_{\bullet})^2 \right]^{1/2} \right. = -1$$

Whenever both $|\rho_{\alpha, \tau}|$ and $|\rho_{\beta, \gamma}|$ are close to 1, then Tukey's model should be an adequate model to use to describe the data.

The cell means given in Table 1.8 were generated by Mandel's model. The values of the parameters in (1.11.1) necessary to describe the data in Table 1.8 are

$$\mu = 29$$

$$\tau' = [-5 \quad 2 \quad 4 \quad 0 \quad -1]$$

$$\beta' = [-10 \quad 4 \quad -2 \quad 8]$$

$$\lambda = 30.3315$$

$$\alpha' = [0 \quad .2236 \quad .6708 \quad -.2236 \quad -.6708]$$

$$\gamma' = [-.7372 \quad .2949 \quad -.1474 \quad .5898]$$

The correlations between the $\alpha$'s and the $\tau$'s and between the $\beta$'s and the $\gamma$'s are $\rho_{\alpha, \tau} = .56$ and $\rho_{\beta, \gamma} = 1$, respectively. Whenever $|\rho_{\beta, \gamma}|$ is close to 1 and $|\rho_{\alpha, \tau}|$ is much smaller than 1 or vice versa, a Mandel bundle-of-straight-lines model will usually be a good model for describing the data.

The preceding discussion makes it apparent that the multiplicative interaction model can be a valuable aid for helping experimenters analyze and interpret data resulting from two-way treatment structure experiments. This is true regardless of whether there are independent replications of the treatment combinations in the experiment or not.

Next consider the model given by

**1.11.3**
**The Real Data Case**

$$y_{ij} = \mu + \tau_i + \beta_j + \lambda\alpha_i\gamma_j + \epsilon_{ij}$$

$$i = 1, 2, \ldots, t, \; j = 1, 2, \ldots, b \qquad (1.11.3)$$

where the parameters on the right-hand side of (1.11.3) satisfy the same restrictions as those given in (1.11.2). Assume the errors $\epsilon_{ij}$ are distributed independently with mean zero and common variance $\sigma^2$. To obtain hypothesis-testing results, also assume the errors are normally distributed. Least squares estimates of the parameters in the model (1.11.3) are

$$\hat{\mu} = \bar{y}_{..}$$

$$\hat{\tau}_i = \bar{y}_{i.} - \bar{y}_{..} \qquad i = 1, 2, \ldots, t$$

$$\hat{\beta}_j = \bar{y}_{.j} - \bar{y}_{..} \qquad j = 1, 2, \ldots, b \qquad (1.11.4)$$

$\hat{\lambda}^2 = $ largest characteristic root of $\mathbf{Z'Z}$ (or $\mathbf{ZZ'}$) where $\mathbf{Z}$ is the $t \times b$ matrix of residuals after fitting an additive model to the data, i.e., $z_{ij} = y_{ij} - \bar{y}_{i.} - \bar{y}_{.j} + \bar{y}_{..}$.

$\hat{\alpha} = $ normalized characteristic vector of $\mathbf{ZZ'}$ corresponding to the largest characteristic root $\hat{\lambda}^2$

$\hat{\gamma} = $ normalized characteristic vector of $\mathbf{Z'Z}$ corresponding to the largest characteristic root $\hat{\lambda}^2$

Before continuing note that:

1. $\mathbf{Z} = (z_{ij})$ where $z_{ij} = y_{ij} - \bar{y}_{i.} - \bar{y}_{.j} + \bar{y}_{..}$ .
2. A normalized vector $\boldsymbol{u}$ is one which satisfies $\boldsymbol{u'u} = 1$; that is, one which has length equal to 1.
3. The estimates given above are determined uniquely except for sign. To obtain the proper sign for $\hat{\lambda}$ for a given $\hat{\alpha}$ and $\hat{\gamma}$, one can take $\hat{\lambda} = \hat{\alpha}'\mathbf{Z}\hat{\gamma}$.
4. When the errors are distributed normally and independently, these estimators are also maximum likelihood estimators in addition to being least squares estimators.

To conclude this section a test of $H_0: \lambda = 0$ versus $H_1: \lambda \neq 0$ in model (1.11.3) is given. Let $\ell_1 > \ell_2 > \cdots > \ell_p$ be the nonzero characteristic roots of $\mathbf{Z'Z}$ (or $\mathbf{ZZ'}$) where $p = \min(t-1, b-1)$. A generalized likelihood ratio test statistic for testing $H_0$ versus $H_1$ is given by

**1.11.4**
**A Test for Interaction**

$$U_1 = \ell_1/\text{RSS}$$

where

$$RSS = \sum_i \sum_j z_{ij}^2$$

is the residual sum of squares from an additive model. The hypothesis $H_0$ is rejected for large values of $U_1$. Tables of 10%, 5%, and 1% critical points for some selected values of $p = \min(t - 1, b - 1)$ and $n = \max(t - 1, b - 1)$ are given in Appendix Table A.1. These critical points have been determined from Schuurman, Krishnaiah, and Chattopadhyah (1973) when possible, and from Johnson and Graybill (1972) for those cases not given by Schuurman et al.

**1.11.5**
**An Example**

As an example consider the data given in Table 1.1. The least squares estimates of $\mu$, $\tau$, and $\beta$ are

$$\hat{\mu} = 25.03$$
$$\hat{\tau}' = [-3.030 \quad -2.755 \quad -.030 \quad 2.420 \quad 3.395]$$
$$\hat{\beta}' = [-12.47 \quad -6.49 \quad 5.15 \quad 13.81]$$

The matrix of residuals from the additive model is

$$\mathbf{Z} = \begin{bmatrix} 2.77 & 4.09 & -1.45 & -5.41 \\ 3.895 & 1.115 & -.425 & -4.585 \\ 5.27 & 1.49 & -3.85 & -2.91 \\ -2.88 & -3.56 & 4.3 & 2.14 \\ -9.055 & -3.135 & 1.425 & 10.765 \end{bmatrix}$$

The largest characteristic root of $\mathbf{Z}'\mathbf{Z}$ (and $\mathbf{Z}\mathbf{Z}'$) is 360.81. The corresponding characteristic vectors of $\mathbf{Z}'\mathbf{Z}$ and $\mathbf{Z}\mathbf{Z}'$ are

$$\hat{\gamma}' = [.6157 \quad .3016 \quad -.2250 \quad -.6923]$$

and

$$\hat{\alpha}' = [-.3691 \quad -.3161 \quad -.3461 \quad .2788 \quad .7525]$$

respectively. The least squares estimate of $\lambda$ is $\hat{\lambda} = -\sqrt{360.81} = -18.995$.

The value of the characteristic root test is $U_1 = 360.81/400.447 = .9010$. The value of the 5% critical point from Table A.1 is .8811. Since $U_1 > .8811$, $H_0: \lambda = 0$ is rejected at a 5% significance level.

Finally, note that the first three elements of $\hat{\alpha}$ are nearly the same. This tends to make one believe that the lower three temperature levels do not interact with humidity levels. Justification for statements such as this will be examined in Chapter 3. In addition, a more complete analysis of data satisfying the multiplicative interaction model is given in that chapter. The multiplicative model is generalized to a model containing additional multiplicative interaction terms in Chapter 2.

To conclude this section, the procedures SAS® GLM and SAS® MATRIX **1.11.6**
are used to fit model (1.11.3) to the data in Table 1.1 and to obtain the   **A SAS Analysis**
characteristic root test for interaction. The required SAS® commands are

```
PROC GLM;
CLASSES TEMP HUMIDITY;
MODEL HEIGHT = TEMP HUMIDITY/SOLUTION;
OUTPUT OUT = RESID     RESIDUAL = RS;

PROC MATRIX FUZZ;
FETCH RR DATA = RESID;
NOTE DATA AND RESIDUALS; PRINT RR;
R = RR(*,4); NO_ROW = 5; NO_COL = 4;
Z = SHAPE(R, NO_COL);
NOTE PAGE ***** RESIDUAL MATRIX Z ******;
PRINT Z;
ZPZ = Z'*Z; ZZP = Z*Z';
EIGEN ROOTS EZPZ ZPZ;
NOTE PAGE EIGENVALUES;
PRINT ROOTS;
NOTE SKIP = 2 EIGENVECTORS OF Z'Z;
PRINT EZPZ;
EIGEN ROOTS EZZP ZZP;
NOTE SKIP = 4 EIGENVECTORS OF ZZ';
PRINT EZZP;
LAMDA = EZZP(*,1)'* Z* EZPZ(*,1);
NOTE SKIP = 2 LAMDA;
L1 = MAX(ROOTS);
RSS = TRACE(ZPZ);

U1 = L1#/RSS;
NOTE SKIP = 2 CHARACTERISTIC ROOT TEST STATISTIC
    FOR H_0:LAMDA = 0;
PRINT U1;
```

The results obtained from the above SAS® MATRIX procedure are
given in Table 1.16.

**Table 1.16** **Results from SAS® MATRIX Giving the Characteristic Root Test for Interaction**

Eigenvalues of Z*Z

| ROOTS | COL1 | COL2 | COL3 | COL4 |
|-------|------|------|------|------|
| ROW1 | 360.81 | 28.4937 | 11.1434 | 0 |

Eigenvectors of Z*Z

| EZPZ | COL1 | COL2 | COL3 | COL4 |
|------|------|------|------|------|
| ROW1 | 0.615683 | 0.00206929 | −0.60904 | 0.5 |
| ROW2 | 0.301634 | 0.381256 | 0.716701 | 0.5 |
| ROW3 | −0.224976 | −0.816631 | 0.180278 | 0.5 |
| ROW4 | −0.692342 | 0.433306 | −0.287939 | 0.5 |

Eigenvectors of ZZ*

| EZZP | COL1 | COL2 | COL3 | COL4 | COL5 |
|------|------|------|------|------|------|
| ROW1 | −0.369092 | 0.0758715 | −0.76108 | −0.527944 | 0.0068₄ |
| ROW2 | −0.316105 | −0.226019 | 0.098784 | 0.0580743 | 0.9142₀ |
| ROW3 | −0.346141 | 0.461242 | 0.598508 | −0.554981 | −0.0350₃ |
| ROW4 | 0.27881 | 0.73951 | 0.191247 | −0.577809 | −0.0703₀ |
| ROW5 | 0.752529 | 0.428416 | −0.127384 | −0.275745 | 0.3973₀ |

Lamda

| LAMDA | COL1 |
|-------|------|
| ROW1 | −18.995 |

Characteristic Root Test Statistic for $H_0$:LAMDA = 0

| U1 | COL1 |
|----|------|
| ROW1 | 0.901018 |

From Table 1.16 one can see that one set of least squares estimates of the interaction parameters in (1.11.3) is:

$$\hat{\gamma}' = [.6157 \quad .3016 \quad -.2250 \quad -.6923]$$

$$\hat{\alpha}' = [-.3691 \quad -.3161 \quad -.3461 \quad .2788 \quad .7525]$$

$$\hat{\lambda} = -18.995$$

In the next chapter, a more general multiplicative interaction model is discussed in detail.

# 2

# The Multiplicative
# Interaction Model

## CHAPTER OUTLINE

In Section 1.11.1 a special case of the multiplicative interaction model was introduced. In addition, a test for interaction, based on this special case, was given for a two-way treatment structure having only one observation per treatment combination. In this chapter this model is generalized and its analysis is discussed. Further analyses of two-way treatment structures having one observation per treatment combination using multiplicative interaction models are described in Chapter 3. Chapter 3 also places special emphasis on the important problem of estimating the experimental error variance. The data in Table 1.1 are used once again to illustrate many of the techniques that are introduced. Methods used to obtain computer analyses are also given.

## 2.1 MODEL DEFINITION AND PARAMETER ESTIMATION

Let $\mu_{ij}$ represent the response expected when treatments $T_i$ and $B_j$ are both applied to a given experimental unit for $i = 1, 2, \ldots, t$ and $j = 1, 2, \ldots, b$.

Using matrix decomposition results, it is possible to show that any $t \times b$ matrix of $\mu_{ij}$'s can always be decomposed into the following form:

$$\mu_{ij} = \mu + \tau_i + \beta_j + \lambda_1 \alpha_{1i} \gamma_{1j}$$

$$+ \lambda_2 \alpha_{2i} \gamma_{2j} + \cdots + \lambda_k \alpha_{ki} \gamma_{kj}$$

$$i = 1, 2, \ldots, t, \quad j = 1, 2, \ldots, b \qquad (2.1.1)$$

where

$$k = \text{rank} \ (\Omega)$$

$$\Omega = (\omega_{ij})$$

and

$$\omega_{ij} = \mu_{ij} - \overline{\mu}_{i\bullet} - \overline{\mu}_{\bullet j} + \overline{\mu}_{\bullet\bullet}$$

In order to reduce the number of many possible sets of values the parameters in (2.1.1) may take on and to simplify the discussion of the multiplicative interaction model, it is assumed without any loss of generality that:

1. $\sum_i \tau_i = \sum_j \beta_j = 0$

2. $|\lambda_1| \geq |\lambda_2| \geq \cdots \geq |\lambda_k|$

3. $\sum_i \alpha_{ri} = \sum_j \gamma_{rj} = 0$ for $r = 1, 2, \ldots, k$

4. $\sum_i \alpha_{ri}^2 = \sum_j \gamma_{rj}^2 = 1$ for $r = 1, 2, \ldots, k$

5. $\sum_i \alpha_{ri} \alpha_{r'i} = \sum_j \gamma_{rj} \gamma_{r'j} = 0$ for $r \neq r' = 1, 2, \ldots, k$ $\qquad (2.1.2)$

**Table 2.1  Cell Mean Parameters with Two Cells Contributing to the Interaction. The Cells Are in Different Rows and Columns**

| $\mu_{ij}$ | $B_1$ | $B_2$ | $B_3$ | $B_4$ |
|---|---|---|---|---|
| $T_1$ | 19 | 26 | 23 | 28 |
| $T_2$ | 26 | 30 | 30 | 35 |
| $T_3$ | 28 | 35 | 38 | 37 |
| $T_4$ | 24 | 31 | 28 | 33 |
| $T_5$ | 23 | 30 | 27 | 32 |

The $\omega_{ij}$'s span the interaction space; that is, any contrast in the $\mu_{ij}$'s which measures interaction can be written as a linear combination of the $\omega_{ij}$'s. It is also true that $k \leq \min(b - 1, t - 1)$.

Next, some situations are described where model (2.1.1) is applicable and other models such as Tukey's model and Mandel's bundle-of-straight-lines model are not. First, consider the set of possible $\mu_{ij}$'s given in Table 2.1.

The cell means in Table 2.1 have two treatment combinations which account for all of the interaction in this two-way table. See if you can find them. One set of values (it will be shown later how such sets of values can be determined) for the parameters in model (2.1.1), which satisfy the restrictions in (2.1.2), is

$$k = 2$$

$$\mu = 29.15$$

$$\tau' = [-5.15 \quad 1.10 \quad 5.35 \quad -.15 \quad -1.15]$$

$$\beta' = [-5.15 \quad 1.25 \quad .05 \quad 3.85]$$

$$\lambda_1 = -4.5217$$

$$\lambda_2 = 2.1804$$

$$\alpha_1' = [-.2619 \quad -.1000 \quad .8856 \quad -.2619 \quad -.2619]$$

$$\gamma_1' = [.3104 \quad .2440 \quad -.8648 \quad .3104]$$

$$\alpha_2' = [-.2545 \quad .8889 \quad -.1254 \quad -.2545 \quad -.2545]$$

$$\gamma_2' = [.3920 \quad -.8309 \quad .0469 \quad .3920]$$

Examination of these parameter values reveals that $\alpha_{11} = \alpha_{14} = \alpha_{15}$ and $\alpha_{21} = \alpha_{24} = \alpha_{25}$. These imply that rows 1, 4, and 5 do not interact with the column treatments. Further examination reveals that $\gamma_{11} = \gamma_{14}$ and $\gamma_{21} = \gamma_{24}$, which imply that column treatments 1 and 4 do not interact with any of the row treatments. Thus, the interaction in these data occurs in rows 2 and 3 with columns 2 and 3 and hence must be caused by at least two of the four treatment combinations: $T_2B_2$, $T_2B_3$, $T_3B_2$, and $T_3B_3$. A closer examination of the data in Table 2.1 with attention now placed on these four treatment combinations and their relationship to the other cell means reveals that just two of the four treatment combinations are causing the interaction, namely $T_2B_2$ and $T_3B_3$. This is because the cell means corresponding to treatment combinations $T_2B_3$ and $T_3B_2$ are additive with respect to all cells in the table except $T_2B_2$ and $T_3B_3$. It should also be pointed out that the fact that the parameter values pointed to the cells that contribute to interaction is not a consequence of which set of possible parameter values one might choose. That is, all possible sets would still have $\alpha_{11} = \alpha_{14} = \alpha_{15}$, $\alpha_{21} = \alpha_{24} = \alpha_{25}$, and $\gamma_{11} = \gamma_{14}$ and $\gamma_{21} = \gamma_{24}$.

As a second example consider the parameter values of the $\mu_{ij}$'s given in Table 2.2 and try to determine which treatment combinations are contributing to the interaction in this table.

One set of values for the parameters in model (2.1.1) which satisfy the restrictions in (2.1.1) is

$$k = 2$$

$$\mu = 28.8$$

$$\tau' = [-4.8 \quad 2.2 \quad 3.2 \quad .2 \quad -.8]$$

$$\beta' = [-2.2 \quad 0.0 \quad -.2 \quad 2.4]$$

$$\lambda_1 = -11.1924$$

$$\lambda_2 = -7.9706$$

$$\alpha_1' = [-.3262 \quad .8071 \quad .1715 \quad -.3262 \quad -.3262]$$

$$\gamma_1' = [-.6120 \quad .3236 \quad -.3452 \quad .6337]$$

$$\alpha_2' = [.1641 \quad .3855 \quad -.8778 \quad .1641 \quad .1641]$$

$$\gamma_2' = [.4324 \quad -.6258 \quad -.3520 \quad .5454]$$

Examination of these parameter values reveals that $\alpha_{11} = \alpha_{14} = \alpha_{15}$ and $\alpha_{21} = \alpha_{24} = \alpha_{25}$, which implies that row treatments 1, 4, and 5

**Table 2.2   Cell Means with Two Rows Containing All of the Interaction**

| $\mu_{ij}$ | $B_1$ | $B_2$ | $B_3$ | $B_4$ |
|---|---|---|---|---|
| $T_1$ | 19 | 26 | 23 | 28 |
| $T_2$ | 33 | 30 | 35 | 26 |
| $T_3$ | 34 | 27 | 30 | 37 |
| $T_4$ | 24 | 31 | 28 | 33 |
| $T_5$ | 23 | 30 | 27 | 32 |

do not interact with any of the column treatments. Since none of the elements in the two $\gamma$ vectors are equal, we conclude that row treatments 2 and 3 interact with all column treatments. Also row treatments 2 and 3 interact between themselves and all of the column treatments since $\alpha_{12} \neq \alpha_{13}$ and $\alpha_{22} \neq \alpha_{23}$. In fact, if row treatments 2 and 3 did not interact with the column treatments and were, instead, additive with the column treatments, i.e., $\alpha_{12} = \alpha_{13}$ and $\alpha_{22} = \alpha_{23}$, then $k$ would have been equal to 1.

As a final example consider the parameter values in Table 2.3, which has one row interacting with the column treatments as well as one column interacting with the row treatments.

The values in the lower right-hand diagonal $4 \times 3$ block of Table 2.3 are completely additive if considered by themselves. Thus, all of the interaction in this set of means can be attributed to the first row treatment interacting with the column treatments and the first column treatment interacting with the row treatments. One set of values of the parameters in the model (2.1.1) which satisfy the restrictions in (2.1.2) is

$$k = 2$$

$$\mu = 29.05$$

$$\tau' = [-6.3 \quad 1.7 \quad 4.45 \quad .7 \quad -.55]$$

$$\beta' = [-3.05 \quad .95 \quad -.65 \quad 2.75]$$

$$\lambda_1 = -10.2582$$

$$\lambda_2 = 3.474$$

$$\alpha_1' = [.8618 \quad -.4029 \quad -.2422 \quad -.1887 \quad -.0280]$$

$$\gamma_1' = [-.5494 \quad .2111 \quad -.3769 \quad .7152]$$

$$\alpha_2' = [-.2395 \quad -.6146 \quad -.0365 \quad .1562 \quad .7343]$$

$$\gamma_2' = [.6694 \quad -.2001 \quad -.6828 \quad .2135]$$

Unfortunately, in this example, examination of these parameter values reveals little information about the nature of the interaction in this set of means. This is because the nature of the interaction in Table

**Table 2.3   One Row and One Column Causing the Interaction**

| $\mu_{ij}$ | $B_1$ | $B_2$ | $B_3$ | $B_4$ |
|------------|-------|-------|-------|-------|
| $T_1$ | 24 | 22 | 26 | 19 |
| $T_2$ | 24 | 33 | 30 | 36 |
| $T_3$ | 29 | 35 | 32 | 38 |
| $T_4$ | 26 | 31 | 28 | 34 |
| $T_5$ | 27 | 29 | 26 | 32 |

2.3 is slightly more complex than it was in the preceding two examples. One of the reasons that these parameter values do not reveal the nature of the interaction is because of the restrictions that were placed on those parameter values by (2.1.2). To illustrate, other sets of parameter values for model (2.1.1) can be obtained for the means in Table 2.3, but the values do not satisfy the restrictions in (2.1.2). Another set of parameter values is given by

$$k = 2$$

$$\mu = 32$$

$$\tau' = [-8.158 \quad 1 \quad 3 \quad -1 \quad -3]$$

$$\beta' = [-4.368 \quad 0 \quad -3 \quad 3]$$

$$\lambda_1 = 7.114$$

$$\lambda_2 = 9.565$$

$$\alpha_1' = [.6362 \quad -.6511 \quad -.2294 \quad -.0888 \quad .3329]$$

$$\gamma_1' = [1 \quad 0 \quad 0 \quad 0]$$

$$\alpha_2' = [1 \quad 0 \quad 0 \quad 0 \quad 0]$$

$$\gamma_2' = [0 \quad -.1926 \quad .5392 \quad -.8198]$$

Even though this set of parameter values does not satisfy the restrictions in (2.1.2), it does reveal the nature of the interaction in Table 2.3. To see this one can examine the interaction matrix given by $\lambda_1 \alpha_1 \gamma_1' + \lambda_2 \alpha_2 \gamma_2'$. Also note that we were only able to obtain this particular solution set because we knew the type of solution that we were looking for. In real data situations, this may not always be possible. At the present time we are conjecturing that some "factor analysis" techniques may be useful in this regard. That is, given a solution set that satisfies the restrictions of (2.1.2), it may be possible to "rotate" that solution into other solutions using "factor analysis" techniques and, as a result, obtain another solution set which does reveal the nature of the interaction. Unfortunately we do not, as yet, know how this can be done.

Next a real data situation is examined. Given a set of experimental data from a two-way treatment situation with one observation per cell, we consider models of the form,

$$y_{ij} = \mu + \tau_i + \beta_j + \lambda_1 \alpha_{1i} \gamma_{1j} + \lambda_2 \alpha_{2i} \gamma_{2j} + \cdots$$

$$+ \lambda_k \alpha_{ki} \gamma_{kj} + \epsilon_{ij}$$

$$i = 1, 2, \ldots, t, \quad j = 1, 2, \ldots, b$$

where it is assumed that the parameters satisfy the restrictions given in (2.1.2) and that the errors are distributed independently with mean zero and common variance $\sigma^2$. One set of least squares estimates of the parameters in this model is

$$\hat{\mu} = \bar{y}_{..}$$

$$\hat{\tau}_i = \bar{y}_{i.} - \bar{y}_{..} \qquad i = 1, 2, \ldots, t$$

$$\hat{\beta}_j = \bar{y}_{.j} - \bar{y}_{..} \qquad j = 1, 2, \ldots, b$$

$$\hat{\lambda}_r^2 = \ell_r \qquad r = 1, 2, \ldots, k$$

where $\ell_1 > \ell_2 > \cdots > \ell_k > \ell_{k+1} > \cdots > \ell_p$ are the nonzero characteristic roots of $\mathbf{Z}'\mathbf{Z}$ (or $\mathbf{ZZ}'$),

$$\mathbf{Z} = (z_{ij})$$

$$z_{ij} = y_{ij} - \bar{y}_{i.} - \bar{y}_{.j} + \bar{y}_{..}$$

$$p = \min \, (b - 1, \, t - 1)$$

$\hat{\boldsymbol{\alpha}}_r$ = normalized characteristic vector of $\mathbf{ZZ}'$ corresponding

to the characteristic root, $\ell_r$, $r = 1, 2, \ldots, k$

$\hat{\boldsymbol{\gamma}}_r$ = normalized characteristic vector of $\mathbf{Z}'\mathbf{Z}$ corresponding

to the characteristic root $\ell_r$, $r = 1, 2, \ldots, k$

The proper sign for $\hat{\lambda}_r$ is obtained by taking

$$\hat{\lambda}_r = \hat{\boldsymbol{\alpha}}_r' \mathbf{Z} \hat{\boldsymbol{\gamma}}_r \qquad r = 1, 2, \ldots, k$$

If, in addition to being independent, the errors are distributed normally, then the above estimators are also maximum likelihood estimators. The estimation of $\sigma^2$ is discussed in Chapter 3.

## 2.2 SELECTING AN APPROPRIATE NUMBER OF TERMS TO DESCRIBE THE INTERACTION

If an experimenter wishes to use a model of the type given in (2.1.1), then it is necessary to determine the number of multiplicative interaction terms necessary to adequately model the data. In trying to make this decision, problems are encountered that are somewhat similar to the ones encountered when building good regression models. The objective is to obtain a simple model (one with as few terms as possible) and, at the same time, obtain an adequate model. The reader should recall that in polynomial regression situations, it is always possible to fit a polynomial model of degree $(n-1)$ to $n$ data points, but such models are generally not good ones. Such models do a good job of predicting the mean response at observed $x$ values, but could be very bad at predicting responses at unobserved $x$ values. See Chapter 7 for more on this subject. A similar situation arises for data from a two-way treatment structure in that it is always possible to take $k = \min \, (t - 1, \, b - 1)$ and fit the data exactly with model (2.1.1), but such a model would not likely be good because one would be overfitting the data. Hence, it is desirable to select a model having as few terms as possible in it and still provide an adequate fit to the data.

Assuming for the moment that we would know how to test the necessary hypotheses, a reasonable model-building procedure might be

1. Test $H_{01}$: $\lambda_1 = 0$ vs. $H_{a1}$: $\lambda_1 \neq 0$ in the model

$$y_{ij} = \mu + \tau_i + \beta_j + \lambda_1 \alpha_{1i} \gamma_{1j} + \epsilon_{ij}$$

2. If one fails to reject $H_{01}$, conclude the data are additive, and complete an analysis of the data accordingly. However, if one rejects $H_{01}$, then test

$$H_{02}: \lambda_2 = 0, \lambda_1 \neq 0 \text{ vs. } H_a: \lambda_2 \neq 0, \lambda_1 \neq 0$$

in the model

$$y_{ij} = \mu + \tau_i + \beta_j + \lambda_1 \alpha_{1i} \gamma_{1j} + \lambda_2 \alpha_{2i} \gamma_{2j} + \epsilon_{ij}$$

3. If one fails to reject $H_{02}$, conclude the appropriate model is the one given in (1), and complete the analysis accordingly. If one rejects $H_{02}$, then test

$$H_{03}: \lambda_3 = 0, \lambda_2 \neq 0, \lambda_1 \neq 0 \text{ vs. } H_{a3}: \lambda_3 \neq 0, \lambda_2 \neq 0, \lambda_1 \neq 0$$

in the model

$$y_{ij} = \mu + \tau_i + \beta_j + \lambda_1 \alpha_{1i} \gamma_{1j} + \lambda_2 \alpha_{2i} \gamma_{2j} + \lambda_3 \alpha_{3i} \gamma_{3j} + \epsilon_{ij}$$

4. Continue in this manner until failing to reject a hypothesis.

One disadvantage of the above procedure is similar to that encountered with a forward selection procedure in multiple regression problems. Namely, one interaction term may account for so little of the interaction variability that it itself appears to be nonsignificant. Thus, it might be preferable to use a procedure similar to the one given above, but with a small variation. One could test successive hypotheses as outlined above and stop only when failing to reject two successive hypotheses, after which it would be concluded that the correct value of $k$ is the value at which the last rejection was made. In our use of the above model in real data situations, an experiment requiring more than two terms has never been encountered, and in most cases only one term has been required.

A likelihood ratio test of $H_{01}$ versus $H_{a1}$ was given in Section 1.11.4. The test was: Reject $H_{01}$ in favor of $H_{a1}$, if

$$U_1 = \ell_1 \bigg/ \sqrt{\sum_{ij} z_{ij}^2} > C_\alpha$$

where $C_\alpha$ is the $\alpha \cdot 100\%$ critical point obtained from Appendix Table A.1 with $p = \min(t - 1, b - 1)$ and $n = \max(t - 1, b - 1)$. Note that $U_1$ is also equal to

$$U_1 = \ell_1/(\ell_1 + \ell_2 + \cdots + \ell_p)$$

A likelihood ratio test statistic for testing $H_{02}$ versus $H_{a2}$ is $U_2 = \ell_2/(\ell_2 + \ell_3 + \cdots + \ell_p)$. Hegemann and Johnson (1976) gave approximate critical points for the distribution of $U_2$, which are repeated in Table A.2. One rejects $H_{02}$ if $U_2$ is greater than the tabled value in Table A.2.

As one might now guess, likelihood ratio test statistics for $H_{0k}$ versus $H_{ak}$, $k = 3, 4, \ldots, p - 1$, are given by

$$U_k = \ell_k/(\ell_k + \ell_{k+1} + \cdots + \ell_p)$$

Yochmowitz and Cornell (1978) recommended using the critical points in Table A.1 with $p = \min(b, t) - k$ and $n = \max(b, t) - k$ for $k = 2, 3, \ldots, p - 1$. For want of anything better, at the present time at least, we will use their suggestion for values of $k \geq 3$.

As an example consider the data in Table 1.1. From Table 1.16 the three nonzero roots of the matrix $\mathbf{Z'Z}$ are $\ell_1 = 360.81$, $\ell_2 = 28.49$, and $\ell_3 = 11.14$. Thus, $U_1 = .901$ and $U_2 = .719$. This value of $U_1$ is significant at the 5% level since $U_1 > .882 = C_{.05}$ from Table A.1. The value of $U_2$ is not significant, since $U_2 < .971$, the 10% critical point from Table A.2. A model with three multiplicative interaction terms would fit the data exactly, so we do not consider testing the significance of $U_3$. From Table 1.16, one can also see that

$$\hat{\lambda}_1 = -18.995$$

$$\hat{\alpha}_1' = [-.369 \quad -.316 \quad -.346 \quad .279 \quad .753]$$

$$\hat{\gamma}_1' = [.616 \quad .302 \quad -.225 \quad -.692]$$

The near equality of the first three elements of the $\hat{\alpha}_1$ vector suggests that there may be no interaction between the column treatments and the first three row treatments. A procedure for testing this conjecture is presented in the next chapter. In addition, the estimation $\sigma^2$ is discussed and other methods of making inferences utilizing the multiplicative interaction model are considered in Chapter 3.

# 3

# More on the Analysis of Multiplicative Interaction Models

## CHAPTER OUTLINE

$\mathbf{I}$n this chapter it is assumed that the multiplicative interaction model with a single interaction component has been determined to adequately describe a given two-way treatment structure data set. Complete analyses of data sets described by such models are discussed in this chapter. The estimation of $\sigma^2$ is discussed in Section 3.1. Tests of hypotheses on row and column main effects are discussed in Section 3.2, and methods which show how to locate subsets of the data table which are free of interaction are given in Section 3.3. The knowledge gained in these three subsections is utilized to provide an improved estimate of $\sigma^2$ in Section 3.4.

Model (2.1.1) with $k = 1$ may be written as

$$y_{ij} = \mu + \tau_i + \beta_j + \lambda \alpha_i \gamma_j + \epsilon_{ij}$$

$$i = 1, 2, \ldots, t, \quad j = 1, 2, \ldots, b \qquad (3.1.1)$$

where $\epsilon_{ij} \sim$ i.i.d. $N(0, \sigma^2)$. Maximum likelihood estimates of $\mu$, $\tau_1, \ldots, \tau_t, \beta_1, \ldots, \beta_b, \lambda, \alpha_1, \ldots, \alpha_t$, and $\gamma_1, \ldots, \gamma_b$ are given in (1.11.4). The maximum likelihood estimate of $\sigma^2$ is $(\sum_{ij} z_{ij}^2 - \ell_1)/bt$. However, this is a biased estimate of $\sigma^2$. A much better estimate of $\sigma^2$ is

$$\hat{\sigma}^2 = \left( \sum_{ij} z_{ij}^2 - \ell_1 \right) / [(b-1)(t-1) - \nu_1] \qquad (3.1.2)$$

where $\nu_1$ is the expected value of $\ell_1/\sigma^2$ when $\lambda = 0$ in (3.1.1). The estimator $\hat{\sigma}^2$ in (3.1.2) is unbiased when $\lambda = 0$ and the results of a simulation study in Hegemann and Johnson (1976) have shown that the estimator in (3.1.2) is not too badly biased when $\lambda \neq 0$. The estimator $\hat{\sigma}^2$ tends to be a conservative estimate of $\sigma^2$; that is, $\hat{\sigma}^2$ tends to overestimate $\sigma^2$ rather than underestimate $\sigma^2$. Although $[(b-1)(t-1) - \nu_1]\hat{\sigma}^2/\sigma^2$ is not distributed as a chi-square distribution, the constant $(b-1)(t-1) - \nu_1$ provides some information as to how good $\hat{\sigma}^2$ might be as an estimate of $\sigma^2$. The larger the value of $(b-1)(t-1) - \nu_1$, the better the estimate. Mandel (1971) and others have referred to $(b-1)(t-1) - \nu_1$ as the pseudo-degrees of freedom corresponding to $\hat{\sigma}^2$. Values for $\nu_1$, which were obtained by simulation, are given in Table A.3, where $p = \min(b, t) - 1$ and $n = \max(b, t) - 1$.

Hegemann and Johnson (1976) showed how to obtain conservative $(1 - \alpha)$ 100% upper confidence bounds on $\sigma^2$.

Main effect hypotheses that can be tested are:

$$H_{01}: \bar{\mu}_{1\bullet} = \bar{\mu}_{2\bullet} = \cdots = \bar{\mu}_{t\bullet}$$

and

$$H_{02}: \bar{\mu}_{\bullet 1} = \bar{\mu}_{\bullet 2} = \cdots = \bar{\mu}_{\bullet b}$$

**51**

Whether or not such hypotheses are reasonable depends upon the nature and extent of the interaction [see Milliken and Johnson (1984), p. 117]; Hegemann and Johnson (1976) showed that a likelihood ratio test statistic for testing $H_{01}$ is

$$F_T = \frac{b \sum_i (\bar{y}_{i\cdot} - \bar{y}_{\cdot\cdot})^2/(t-1)}{\hat{\sigma}^2}$$

and a likelihood ratio test statistic for testing $H_{02}$ is

$$F_B = \frac{t \sum_j (\bar{y}_{\cdot j} - \bar{y}_{\cdot\cdot})^2/(b-1)}{\hat{\sigma}^2}$$

One rejects $H_{01}$ for large values of $F_T$ and rejects $H_{02}$ for large values of $F_B$. The sampling distributions of these two statistics are not $F$ distributions. Hegemann and Johnson (1976) studied the sampling distributions of $F_T$ and $F_B$ when $\lambda \neq 0$. They discovered that the critical points for $F_T$ and $F_B$ are generally decreasing as $\lambda$ increases, and hence critical points for $F_T$ and $F_B$ when $\lambda = 0$ can be used for conservative tests of $H_{01}$ and $H_{02}$ when $\lambda \neq 0$. Tables of such critical points were given by Hegemann and Johnson; they are reproduced in Appendix Table A.4.

**3.3 INFERENCES ON THE INTERAC-TION PARAM-ETERS**

Once an experimenter has determined there is interaction in the data, an attempt should be made to determine where it exists and whether it is possible that a few selected treatment combinations are responsible for most of the interaction. Two important consequences of knowing where interaction exists in the data are

1. Portions of the data that are free from interaction could be used to obtain a more conventional type of estimator of $\sigma^2$ (this is illustrated later).

2. One can obtain tests for "main effects" in subtables which are free of interaction, and such main effect hypotheses are easily interpreted.

The multiplicative interaction model with a single interaction term can be written as

$$y_{ij} = \mu + \tau_i + \beta_j + \lambda \alpha_i \gamma_j + \epsilon_{ij}$$
$$i = 1, 2, \ldots, t, \quad j = 1, 2, \ldots, b \tag{3.3.1}$$

where it is assumed that

$$\sum_i \tau_i = \sum_j \beta_j = \sum_i \alpha_i = \sum_j \gamma_j = 0$$

and

$$\sum_i \alpha_i^2 = \sum_j \gamma_j^2 = 1$$

Recall from Section 1.11.1, equality of some of the $\alpha_i$'s in (3.3.1) does, in fact, imply that the corresponding rows do not interact with the column treatments. Similarly, equality of some of the $\gamma_j$'s implies that the corresponding columns do not interact with the row treatments.

To determine whether some rows (or columns) are free of interaction, one would like to test hypotheses of the form $H_0$: $\alpha_1 = \alpha_2 = \cdots = \alpha_r$ (or $H_0$: $\gamma_1 = \gamma_2 = \cdots = \gamma_s$). Marasinghe and Johnson (1981) gave a likelihood ratio test of $H_0$: $\mathbf{H\alpha} = \mathbf{0}$ in model (3.3.1) where $\mathbf{H}$ is any $q \times t$ matrix of row contrasts and $\mathbf{\alpha'} = [\alpha_1 \alpha_2 \ldots \alpha_t]$. The procedure can also be used to test $\mathbf{G\gamma} = 0$, where $\mathbf{\gamma'} = [\gamma_1 \gamma_2 \ldots \gamma_b]$, by merely interchanging rows and columns in the analysis.

Marasinghe and Johnson (1981) showed that the likelihood ratio test of $H_0$: $\mathbf{H\alpha} = \mathbf{0}$ in model (3.3.1) is of the form: Reject $H_0$ if

$$\Lambda = \left[ \sum_{ij} z_{ij}^2 - \ell_1 \right] \Big/ \left[ \sum_{ij} z_{ij}^2 - \ell_1^* \right] < C_\alpha \qquad (3.3.2)$$

where $C_\alpha$ satisfies $P_{H_0}(\Lambda \leq C_\alpha) = \alpha$ and $\ell_1^*$ is $Ch_{\max}[\mathbf{Z'(I-H'(HH')^{-1}H)Z}]$, i.e., $\ell_1^*$ is the largest characteristic root of $\mathbf{Z'(I - H'(HH')^{-1}H)Z}$ ($\mathbf{Z}$ is defined in 1.11.3).

Marasinghe and Johnson (1981) also showed that the null distribution of the above likelihood ratio statistic depends on $\delta = \lambda/\sigma$ and $q$ as well as $b$ and $t$. Marasinghe and Johnson obtained critical points $C_\alpha$ for different values of $\delta$ when $q = 1$ or 2.

The analysis of real data is complicated by the fact that $\delta$ is unknown. To overcome this difficulty, Marasinghe and Johnson recommended that one estimate $\delta$ by

$$\tilde{\delta} = (\ell_1/\hat{\sigma}^2)^{1/2} \qquad (3.3.3)$$

where $\hat{\sigma}^2$ is given by (3.1.2).

If it has been determined that model (3.3.1) is an adequate model, they recommend that one proceed as follows: Take $q = 1$ and choose any contrast of the $\alpha$'s to be tested equal to zero. Compute the corresponding test statistic given by (3.3.2). Compute the estimate of $\delta$ given by (3.3.3) and round the estimate to the next lowest integer. Use this value of $\delta$ along with the known values of $t$ and $b$ to obtain the appropriate critical point from Appendix Table A.5, A.6, or A.7, which are reproduced from Marasinghe and Johnson. If $\Lambda$ is smaller than the tabulated critical point, conclude that the chosen contrast is significantly different from zero.

This recommended procedure is philosophically equivalent to Fisher's LSD procedure discussed in Section 3.4 of Milliken and Johnson (1984). That is, contrasts of the $\alpha$'s are examined only if the characteristic root test has previously shown that there is significant interaction in the data.

To illustrate the described procedure, consider the set of data in Table 3.1. Davies (1954) used this set of data to demonstrate the analysis

**Table 3.1   Data from Davies (1954, p. 305). Factor A Is
Quantitative and Factor B Is Qualitative**

| Level of Factor A | LEVEL OF FACTOR B | | | |
|---|---|---|---|---|
| | *1* | *2* | *3* | *4* |
| 1 | 28.2 | 23.5 | 17.4 | 10.1 |
| 2 | 29.3 | 24.8 | 15.2 | 11.5 |
| 3 | 33.7 | 24.1 | 17.8 | 15.6 |
| 4 | 41.2 | 34.7 | 14.7 | 9.9 |
| 5 | 50.9 | 32.8 | 16.6 | 4.7 |

of two-way factorial experiments. Davies used an additive model to analyze the data. He obtained 33.97 with 12 degrees as an estimate of $\sigma^2$. Davies did not test for interaction. For these data the trace of the matrix $\mathbf{Z'Z}$ (the residual sum of squares of the additive model) is 407.656, and the largest characteristic root of $\mathbf{Z'Z}$, $\ell_1$, is 366.317. Since $\ell_1/\text{tr}(\mathbf{Z'Z}) = .899$ is significant at the 5% level (the 5% critical point from Table A.1 is .882), $H_0: \lambda = 0$ is rejected. Thus, the characteristic root test shows that significant interaction is present in these data. The value of the second largest root of $\mathbf{Z'Z}$ is 29.868. The test statistic for testing $\lambda_2 = 0$, $\lambda_1 \neq 0$ in model (2.1.1) with $k = 2$ is $U_2 = .72$. Since this is not significant (the 10% critical point from Table A.2 is .971), we conclude that only one multiplicative interaction term is needed in the model. The maximum likelihood estimates of the interaction parameters in the model are given by

$$\hat{\lambda} = 19.14$$

$$\hat{\boldsymbol{\alpha}} = [-.37, \quad -.32, \quad -.34, \quad .29, \quad .75]'$$

$$\hat{\boldsymbol{\gamma}} = [.69, \quad .23, \quad -.31, \quad -.61]'$$

The estimate of $\sigma^2$ obtained by using $\hat{\sigma}^2$ in (3.1.2) is

$$\hat{\sigma}^2 = (407.65 - 366.31)/(12 - 8.33)$$

$$= 11.264$$

For these data $\hat{\sigma}^2$ is considerably smaller than the estimate of $\sigma^2$ given by Davies in the original analysis, which was 33.97.

An interesting observation from the above analysis is that $\hat{\alpha}_1$, $\hat{\alpha}_2$, and $\hat{\alpha}_3$ are approximately equal. One might conjecture that these estimates are, in fact, estimating the same parameter, and from the remarks made earlier in this context, one is led to suspect that there may be no interaction in the first three rows of these data. The results in this section can be used to formally investigate whether this is truly the case. This is first done by testing hypotheses of rank one ($q = 1$). That is, the equality of the $\alpha$ parameters is tested in a pairwise manner.

The test of $H_{01}$: $\alpha_1 = \alpha_2$ versus $H_{a1}$: $\alpha_1 \neq \alpha_2$ is now considered in detail.

The value of $\nu_1$ needed for estimating $\sigma^2$ from Table A.3 is $\nu_1 = 8.33$, since $p = 3$ and $n = 4$, and hence,

$$\hat{\delta}^2 = \frac{366.317}{41.34/(12 - 8.33)} = 32.475$$

which gives 5.7 as the estimate of $\delta$.

To calculate the observed value of the statistic $\Lambda$ for testing $H_{01}$: $\alpha_1 = \alpha_2$, the largest characteristic root of the matrix $\mathbf{Z}'(\mathbf{I} - \mathbf{H}'(\mathbf{HH}')^{-1}\mathbf{H})\mathbf{Z}$ has to be obtained, where $\mathbf{H} = \begin{bmatrix} 1 & -1 & 0 & 0 & 0 \end{bmatrix}$. This value is determined to be 365.92, and hence

$$\Lambda = \frac{407.65 - 366.317}{407.65 - 365.92} = .9905$$

From Tables A.5, A.6, and A.7, the 1%, 5%, and 10% critical points for $t = 5$, $b = 4$, $q = 1$, and $\delta = 5$ are .2982, .4910, and .6027, respectively. Thus, there is not sufficient evidence to reject the hypothesis $H_{01}$: $\alpha_1 = \alpha_2$ at the 10% significance level (remember one rejects for small values of $\Lambda$). This would indicate that the first two row treatments do not interact with the column treatments. Other pairwise hypotheses of interest are: $H_{02}$: $\alpha_1 = \alpha_3$, $H_{03}$: $\alpha_1 = \alpha_4$, $H_{04}$: $\alpha_1 = \alpha_5$, $H_{05}$: $\alpha_2 = \alpha_3$, $H_{06}$: $\alpha_2 = \alpha_4$, $H_{07}$: $\alpha_2 = \alpha_5$, $H_{08}$: $\alpha_3 = \alpha_4$, $H_{09}$: $\alpha_3 = \alpha_5$, and $H_{010}$: $\alpha_4 = \alpha_5$. Values of $\Lambda$ for these hypotheses are, respectively, .9973, .3580, .1571, .9980, .3837, .1707, .3813, .1624, and .5308. Thus $H_{04}$, $H_{07}$, and $H_{09}$ are rejected at the 1% level, $H_{03}$, $H_{06}$, and $H_{08}$ are rejected at the 5% level, and $H_{010}$ is rejected at the 10% level. Further examination of these results shows that all hypotheses involving either row 4 or 5 show significant interactions, while those involving only row 1, 2, or 3 are not significant. This indicates that all of the interaction in the data can be accounted for by row treatments 4 and 5, and that portion of the data involving only the first three rows is additive.

In the above analysis, the procedure described for $q = 1$ is applied 10 times. The reader is reminded that it was first shown that there was interaction in the data, and the repeated testing was done with $q = 1$ to determine which treatment combinations interact. The first test affords the experimenter protection at approximately the 5% level.

It is interesting to note that the estimate of $\sigma^2$ obtained by fitting an additive model to the first three rows of the data is 2.803 with 6 degrees of freedom. This estimate of $\sigma^2$ and some others will be examined in the next section.

Before concluding this example, it is informative to consider some alternative ways of analyzing these data. One alternative is to test hypotheses of the form $H_0$: $\alpha_i = \alpha_j = \alpha_k$ for $i \neq j \neq k$. The procedure is the same as for $q = 1$ except that the appropriate hypothesis matrices are now of rank 2 and hence the tables of critical points corresponding to $q = 2$ must be used. Such tables are given in Tables A.8, A.9, and A.10. From these tables the 1%, 5%, and 10% critical points for $t = 5$, $b = 4$, $q = 2$, and $\delta = 5$ are .2104, .3604, and .4548, respectively. The value of $\Lambda$ for testing $\alpha_1 = \alpha_2 = \alpha_3$ is .9905, which is not significant. This is additional support for the claim that there is no interaction in the first three rows of this data set. The values of $\Lambda$ for comparing all other subsets of three rows ranged from .1236 to .3083. All of these are significant at the 5% level, indicating that there is no other subset of three rows which is free of interaction.

Another possible method of analysis is to perform Tukey's single-degree-of-freedom test for interaction to two rows at a time. This was done for all possible pairs of rows in the above example. The values of Tukey's $F$ statistics for testing for nonadditivity in the pairs of rows (1,2), (1,3), . . . , (4,5) are .122, .118, 82.2, 68.8, .0085, 151.0, 72.3, 8.55, 69.7, and 2.89, respectively. Once again, one finds that those hypotheses involving only row 1, 2, or 3 would not be rejected. The only discrepancy between these results and those obtained earlier is that the last hypothesis is not rejected using Tukey's test, while the corresponding hypothesis $H_{010}$ was rejected using the earlier procedure.

This alternative, based on Tukey's test, seems to be a viable one. This is true for these data, but it cannot be expected to hold for all data sets.

**3.4 IMPROVED ESTIMATION OF $\sigma^2$ IN THE MULTIPLICATIVE INTERACTION MODEL**

In Section 3.1 an estimator of the error variance $\sigma^2$ was proposed. In Section 3.3 a test procedure was given which enables us to find rows (or columns) in the data matrix which may be additive. In this subsection an improved estimator of $\sigma^2$ is given which makes use of the additional information obtained about those portions of the data which are additive.

In Chapter 1 several tests for nonadditivity in two-way cross-classified data with one observation per treatment combination were given. While the ability to determine whether a particular data set is additive is useful to the data analyst in many ways, it is also important, if not more important, to be able to obtain a reliable estimator of the experimental error variance. Tests of significance and confidence interval construction when a set of data is determined to be nonadditive require accurate estimation of the experimental error variance.

Consider the multiplicative interaction model given in (3.1.2). When the interaction term is present in this model (that is, when $\lambda \neq 0$), it was proposed that $\sigma^2$ be estimated by (3.3.3). Now suppose it is possible to determine row treatments or column treatments which do not contribute to the interaction effect by using the test statistic given in Section 3.3, which assumes that model (3.3.1) has been determined to adequately fit the data. Let the matrices $\mathbf{H}$ and $\mathbf{Z}$ be as defined in Section 3.3. Then the matrix $\mathbf{Z}'\mathbf{Z}$ can be expressed as the sum of the two matrices $\mathbf{Z}'\mathbf{H}'(\mathbf{HH}')^{-1}\mathbf{HZ}$ and $\mathbf{Z}'(\mathbf{I} - \mathbf{H}'(\mathbf{HH}')^{-1}\mathbf{H})\mathbf{Z}$. When $\mathbf{H\alpha} = 0$, an unbiased estimator of $\sigma^2$ can be obtained from $\mathbf{Z}'\mathbf{H}'(\mathbf{HH}')^{-1}\mathbf{HZ}$, and is given by

$$\hat{\sigma}_1^2 = \mathrm{tr}(\mathbf{Z}'\mathbf{H}'(\mathbf{HH}')^{-1}\mathbf{HZ})/q(b - 1) \qquad (3.4.1)$$

This follows since it can be shown that $\mathrm{tr}(\mathbf{Z}'\mathbf{H}'(\mathbf{HH}')^{-1}\mathbf{HZ})/\sigma^2$ is distributed as $\chi^2[q(b - 1)]$ under $H_0$: $\mathbf{H\alpha} = 0$. This estimator has merit, particularly since it could be used to obtain exact $F$-tests for hypotheses about the treatment effects. However, there is also information about $\sigma^2$ in the matrix $\mathbf{Z}'(\mathbf{I} - \mathbf{H}'(\mathbf{HH}')^{-1}\mathbf{H})\mathbf{Z}$ that could also be utilized. The estimator proposed in this section is a consequence of the following result.

The nonzero characteristic roots of $\mathbf{Z}'(\mathbf{I} - \mathbf{H}'(\mathbf{HH}')^{-1}\mathbf{H})\mathbf{Z}$ are distributed the same as those of a matrix $\mathbf{W}^*$, which is distributed as the Wishart distribution $W_{b-1}(t - q - 1, \sigma^2\mathbf{I}_{b-1}, \mathbf{M}^*)$ where rank $(\mathbf{M}^*) = 1$, and the single nonzero characteristic root of $\mathbf{M}^*$ is $\lambda^2$ when $\mathbf{H\alpha} = \mathbf{0}$. The proof of this theorem follows along the same lines as the proofs of Theorems 4.1 and 4.2 in Johnson and Graybill (1972).

This result enables one to arrive at an estimator of $\sigma^2$ using the characteristic roots of $\mathbf{Z}'(\mathbf{I} - \mathbf{H}'(\mathbf{HH}')^{-1}\mathbf{H})\mathbf{Z}$. This estimator is

$$\hat{\sigma}_2^2 = \frac{\mathrm{tr}[\mathbf{Z}'(\mathbf{I} - \mathbf{H}'(\mathbf{HH}')^{-1}\mathbf{H})\mathbf{Z}] - \ell_1^*}{(b - 1)(t - q - 1) - \nu_1^*} \qquad (3.4.2)$$

where $\ell_1^*$ is defined in (3.3.2) and $\nu_1^* = E_{\lambda=0}(\ell_1^*/\sigma^2)$. It is clear that this estimator is of the same type as $\hat{\sigma}^2$ defined in (3.1.2), the estimator obtained from the characteristic roots of the matrix $\mathbf{Z}'\mathbf{Z}$. Since the distributions of $\mathbf{W}$ and $\mathbf{W}^*$ are similar, the dependence of $\hat{\sigma}_2^2$ on nonzero values of $\lambda$ will be similar to the dependence of $\hat{\sigma}^2$ on nonzero values of $\lambda$. Hence, remarks made previously about $\hat{\sigma}^2$ and its desirability as an estimator of $\sigma^2$ are also applicable to $\hat{\sigma}_2^2$. Values of $\nu_1^*$ can be obtained from those tabulated for $\nu_1$. The values of $n$ and $p$ that are used to determine $\nu_1$ in Table A.3 must be replaced by $\max(b - 1, t - q - 1)$ and $\min(b - 1, t - q - 1)$, respectively, in order to correctly determine $\nu_1^*$.

Marasinghe and Johnson (1981) showed that the characteristic roots of $\mathbf{Z}'\mathbf{H}'(\mathbf{HH}')^{-1}\mathbf{HZ}$ and $\mathbf{Z}'(\mathbf{I} - \mathbf{H}'(\mathbf{HH}')^{-1}\mathbf{H})\mathbf{Z}$ are independently distributed. Thus, a new estimator of $\sigma^2$ can be obtained by taking a weighted average of the two estimators, $\hat{\sigma}_1^2$ and $\hat{\sigma}_2^2$. This pooled estimator, with weights given by the pseudo-degrees of freedom of each estimator, is

$$\hat{\sigma}_p^2 = \frac{q(b - 1)\hat{\sigma}_1^2 + [(b - 1)(t - q - 1) - \nu_1^*]\hat{\sigma}_2^2}{(b - 1)(t - 1) - \nu_1^*}$$

which turns out to be algebraically equivalent to

$$\hat{\sigma}_p^2 = \frac{\mathrm{tr}(\mathbf{Z}'\mathbf{Z}) - \ell_1^*}{(b-1)(t-1) - \nu_1^*} \quad \text{when } t - q > 2 \qquad (3.4.3)$$

The pseudo-degrees of freedom of this estimator are $(b-1)(t-1) - \nu_1^*$.

Note that the numerator of (3.4.3) is also the residual sum of squares of the model (3.3.1) restricted by the null hypothesis, $\mathbf{H\alpha} = \mathbf{0}$. The estimator $\hat{\sigma}_p^2$ should be "better" than $\hat{\sigma}^2$ when $\mathbf{H\alpha} = \mathbf{0}$ is true, in the sense that it will be less likely to overestimate $\sigma^2$, and also in the sense that it will have more pseudo-degrees of freedom than the estimator $\hat{\sigma}^2$. If the characteristic root test rejects $\lambda = 0$ in model (3.3.1) and if the hypothesis $\mathbf{H\alpha} = \mathbf{0}$ is accepted, then $\hat{\sigma}_p^2$ should be used as an estimator of $\sigma^2$.

To illustrate the procedures described above, consider the set of data in Table 3.1. This discussion supplements that given in Section 3.3. There we concluded that there is no significant nonadditivity in the first three rows of the data matrix. One can immediately conjecture that the residual mean square obtained after fitting an additive model to the first three rows of the data matrix should be a good estimator of $\sigma^2$. Computationally this is equal to $\hat{\sigma}_1^2$ defined by (3.4.1). For the data in Table 3.1, $\hat{\sigma}_1^2 = 2.803$ with 6 degrees of freedom. This estimator receives favorable attention because of the fact that if there is no interaction in the first three rows, then $q(b-1)\hat{\sigma}_1^2/\sigma^2$ is distributed as a central chi-square random variable with $q(b-1)$ degrees of freedom, and hence, exact tests and confidence intervals using $\hat{\sigma}_1^2$ could be obtained.

The value of the estimator $\hat{\sigma}_p^2$ defined by (3.4.3) is 5.96 and has 7 pseudo-degrees of freedom as is now shown. This result is obtained by determining both the trace and the largest characteristic root of the matrix $\mathbf{Z}'(\mathbf{I} - \mathbf{H}'(\mathbf{H}\mathbf{H}')^{-1}\mathbf{H})\mathbf{Z}$ where $\mathbf{H}$ is a $2 \times 5$ contrast matrix appropriate for testing $\alpha_1 = \alpha_2 = \alpha_3$. One such $\mathbf{H}$ is

$$\mathbf{H} = \begin{bmatrix} 1 & -1 & 0 & 0 & 0 \\ 1 & 0 & -1 & 0 & 0 \end{bmatrix}$$

The value of 5 for $\nu_1^*$ is obtained from Table A.3 using $p = 2$ and $n = 3$. Hence,

$$\hat{\sigma}_p^2 = \frac{\mathrm{tr}(\mathbf{Z}'\mathbf{Z}) - \ell_1^*}{(b-1)(t-1) - \nu_1^*}$$

$$= \frac{407.656 - 365.919}{12 - 5} = \frac{41.737}{7} = 5.96$$

This improved estimator will help the experimenter to make more accurate decisions concerning the effects of the various treatment combinations.

### Table 3.2   Pairwise Differences from Table 3.1

| | LEVEL OF FACTOR B | | | |
| --- | --- | --- | --- | --- |
| *Level of Factor A* | *1* | *2* | *3* | *4* |
| 1 vs. 4 | $-13.0^*$ | $-11.2^*$ | 2.7 | 0.2 |
| 1 vs. 5 | $-22.7^*$ | $-9.3^*$ | 0.8 | 5.4 |
| 2 vs. 4 | $-11.9^*$ | $-9.9^*$ | 0.5 | 1.6 |
| 2 vs. 5 | $-21.6^*$ | $-8.0$ | $-1.4$ | 6.8 |
| 3 vs. 4 | $-7.5$ | $-10.6^*$ | 3.1 | 5.7 |
| 3 vs. 5 | $-17.2^*$ | $-8.7^*$ | 1.2 | $10.9^*$ |
| 4 vs. 5 | $-9.7^*$ | 1.9 | $-1.9$ | 5.2 |

* Indicates a significant difference.

To illustrate, consider again the data in Table 3.1. To compare the first three levels of factor $A$, it is recommended that one average across the levels of factor $B$ since these levels of factor $A$ do not interact with the levels of factor $B$. One gets $\hat{\mu}_{1\bullet} = 19.8$, $\hat{\mu}_{2\bullet} = 20.2$, and $\hat{\mu}_{3\bullet} = 22.8$, respectively. An approximate 5% LSD value for comparing these three row means to each other is

$$\text{LSD}_{.05} = t_{.025,7} \cdot \hat{\sigma}_p \sqrt{2/b} = 2.365 \cdot 2.441 \cdot .707 = 4.08$$

Hence there is no significant difference between any of the first three levels of factor $A$. Comparing these three levels of factor $A$ to levels 4 and 5 of factor $A$ or comparing levels 4 and 5 to each other requires a different LSD value, because each of these comparisons must be done separately for each level of factor $B$. The required 5% LSD value is

$$\text{LSD}_{.05} = t_{.025,7} \cdot \hat{\sigma}_p \sqrt{2} = 2.365 \cdot 2.441 \cdot 1.414 = 8.16$$

Table 3.2 shows those comparisons which require one to use this LSD value and also indicates those comparisons which are significant.

In comparing the levels of factor $B$ to each other, one can (1) compare the levels of factor $B$ averaged across the first three levels of factor $A$ and (2) compare the levels of factor $B$ separately for levels 4 and 5 of factor $A$. The 5% LSD for the comparisons in (1) is

$$\text{LSD}_{.05} = t_{.025,7} \cdot \hat{\sigma}_p \sqrt{2/3} = 2.365 \cdot 2.441 \cdot .816 = 4.71$$

and the 5% LSD for the comparisons in (2) is, once again,

$$\text{LSD}_{.05} = t_{.025,7} \cdot \hat{\sigma}_p \sqrt{2} = 2.365 \cdot 2.441 \cdot 1.414 = 8.16$$

**3.5
SIMULTA-
NEOUS TESTS
ON THE α's
AND γ's**

In Section 3.3 a procedure was given for testing $H_0$: $\mathbf{H}\alpha = \mathbf{0}$ where $\mathbf{H}$ is a $q \times t$ contrast matrix of rank $q$. The same procedure can also be used to test similar hypotheses about the $\gamma$'s. In this section the likelihood ratio statistic for testing $H_0$: $\mathbf{H}\alpha = \mathbf{0}$ and $\mathbf{G}\gamma = \mathbf{0}$, simultaneously, is given where $\mathbf{H}$ and $\mathbf{G}$ are $q \times t$ and $r \times b$ contrast matrices of ranks $q$ and $r$, respectively. This test can be used to test for the equality of subsets of the $\alpha$'s and the $\gamma$'s simultaneously. Once such a hypothesis is tested, the experimenter is in a better position to be able to reexamine the treatment structure and the data in order to determine whether the interaction pattern exhibited by the data is actually caused by the interacting of two factor levels or is perhaps caused by an outlier. The experimenter may use this additional information to obtain a more reliable estimate of the experimental error variance. A likelihood ratio test statistic $\Lambda$ for testing $H_0$: $\mathbf{H}\alpha = \mathbf{0}$ and $\mathbf{G}\gamma = \mathbf{0}$ vs. $H_a$: $\mathbf{H}\alpha \neq \mathbf{0}$ or $\mathbf{G}\gamma \neq \mathbf{0}$ is given by

$$\Lambda = \left[ \frac{\sum\sum z_{ij}^2 - \ell_1}{\sum\sum z_{ij}^2 - \ell_1^{**}} \right] \tag{3.5.1}$$

where $\ell_1^{**}$ is the largest characteristic root of

$$[\mathbf{I} - \mathbf{H}'(\mathbf{HH}')^{-1}\mathbf{H}]\mathbf{Z}[\mathbf{I} - \mathbf{G}'(\mathbf{GG}')^{-1}\mathbf{G}]\mathbf{Z}'[\mathbf{I} - \mathbf{H}'(\mathbf{HH}')^{-1}\mathbf{H}]$$

The null hypothesis is rejected for small values of $\Lambda$.

The null distribution of $\Lambda$ depends on $b$, $t$, $q$, $r$, and $\delta = \lambda/\sigma$. Marasinghe and Johnson (1982) obtained approximations to the critical points of $\Lambda$, which are reproduced in Tables A.11, A.12, and A.13. Tables are given for $q = 1, 2$ and $r = 1, 2$ for selected values of $b$ and $t$. Only those critical points corresponding to values of $\delta \leq 10$ are included since there is little change in the critical points for values of $\delta > 10$. That is, the critical points tabulated for $\delta = 10$ can be used for all $\delta > 10$. Unfortunately, $\delta$ is an unknown parameter. As before, to overcome this difficulty it is recommended that $\delta$ be estimated by $\hat{\delta} = \sqrt{\ell_1/\hat{\sigma}^2}$. For an illustration of the procedure, consider the set of data from Black (1970) reproduced in Table 3.3. The responses are yields in kilograms per hectare of a spring wheat from an experi-

**Table 3.3   Yields in kg/ha of Spring Wheat**

| Nitrogen in kg/ha | PHOSPHORUS IN KG/HA | | | | |
|---|---|---|---|---|---|
| | 0 | 22 | 45 | 90 | 180 |
| 0 | 1984 | 2550 | 2706 | 2740 | 2954 |
| 45 | 1776 | 2843 | 3306 | 3305 | 3386 |
| 90 | 1797 | 2761 | 3240 | 3227 | 3332 |

ment involving a two-way treatment structure. The treatments consisted of three levels of nitrogen and five levels of phosphorus. An additive two-way model is fit to these data from which the $3 \times 5$ residual matrix $\mathbf{Z}$ is formed. The residual sum of squares from the additive model, which is also the trace of $\mathbf{Z'Z}$, is 258,651.73, and the largest characteristic root of $\mathbf{Z'Z}$ is 257,721.16. Thus, $\ell_1/\text{tr}(\mathbf{Z'Z}) = .9964$. Using the test in Section 1.11, the hypothesis $H_0$: $\lambda = 0$ is rejected at the 1% level. The test shows significant interaction in the data. The estimates of the multiplicative interaction parameters are

$$\hat{\lambda} = 507.66$$
$$\hat{\boldsymbol{\alpha}}' = (.8123, \quad -.4776, \quad -.3347)$$
$$\hat{\boldsymbol{\gamma}}' = (.8231, \quad .0894, \quad -.4124, \quad -.3485, \quad -.1515)$$

and

$$\hat{\sigma}^2 = 930.57/(8 - 6.36) = 567.42$$

As an example of the test proposed, consider testing the hypothesis $H_0$: $\alpha_2 = \alpha_3$ and $\gamma_3 = \gamma_4 = \gamma_5$. Note that $H_0$ being true implies that the last two rows of the data set as well as the last three columns of the data set are free of interactions. Since it has already been determined that there is interaction in the data, acceptance of $H_0$ would imply that the interaction in these data is being caused by either one or both of the first two observations in the first row.

The estimated value of $\delta$ necessary to obtain the appropriate critical point for the test is

$$\hat{\delta} = \sqrt{257,721.16/567.42} = 21.3$$

Thus, the critical point for $\delta = 10$ is used. From Tables A.11 to A.13 the 1%, 5%, and 10% critical points for $t = 3$, $b = 5$, $q = 1$, and $r = 2$ are .0322, .0957, and .1541, respectively.

To compute the observed value of the statistic $\Lambda$, $\ell_1^{**}$ is required. In the present example the hypothesis being tested is $H_0$: $\mathbf{H\alpha} = \mathbf{0}$ and $\mathbf{G\gamma} = \mathbf{0}$ where

$$\mathbf{H} = \begin{bmatrix} 0 & 1 & -1 \end{bmatrix} \quad \text{and} \quad \mathbf{G} = \begin{bmatrix} 0 & 0 & 1 & -1 & 0 \\ 0 & 0 & 1 & 0 & -1 \end{bmatrix}$$

Using these matrices, the characteristic root $\ell_1^{**}$ is determined to be 245,640, and thus,

$$\Lambda = \frac{258,652 - 257,721}{258,652 - 245,640} = .0715$$

Hence, $H_0$ is rejected at the 5% significance level. Johnson and Graybill (1972a) analyzed this same set of data by examining all of the two-by-two contrasts in the experiment. They had suggested that the interaction in these data is due to the observation in the $(1,1)$ cell. Since $H_0$: $\alpha_2 = \alpha_3$ and $\gamma_3 = \gamma_4 = \gamma_5$ was rejected, it appears that there is significant interaction elsewhere in the data set.

To obtain a more complete analysis of this set of data, it was decided to test all possible hypotheses of the form $H_0(i, i'; j, j')$: $\alpha_i = \alpha_{i'}$ and $\gamma_j = \gamma_{j'}$. The 1%, 5%, and 10% critical points for testing hypotheses of this form are .0421, .1278, and .2062, respectively. The values of the test statistic $\Lambda$ ranged from .0038 to .2263. Twenty-six of the 30 possible hypotheses are rejected at the 1% significance level. Three of the remaining four hypotheses are rejected at the 5% level. The only hypothesis accepted is $H_0(2,3;3,4)$: $\alpha_2 = \alpha_3$ and $\gamma_3 = \gamma_4$. The value of $\Lambda$ for testing this hypothesis is .2263. The acceptance of this hypothesis and the rejection of the other 29 hypotheses implies that the observations in the $(1,1)$, $(1,2)$, and $(1,5)$ cells all contribute to interaction with the remaining cells showing no significant nonadditivity.

Given this additional information one can now reestimate $\sigma^2$. The easiest way to do this is to temporarily delete the $(1,1)$, $(1,2)$, and $(1,5)$ observations from the data set and fit an additive two-way model to the remaining data. Doing this, one obtains $\hat{\sigma}^2 = 822.55$ with 5 degrees of freedom.

In the previous section an estimator of $\sigma^2$ was given based on the information discovered about the data set being analyzed. Next, an estimator of $\sigma^2$ is given for the case when we fail to reject the hypothesis $H_0$: $\mathbf{H\alpha} = \mathbf{0}$ and $\mathbf{G\gamma} = \mathbf{0}$. The estimate is

$$\hat{\sigma}_p^2 = \{ \operatorname{tr}(\mathbf{Z'Z}) - \operatorname{tr}[(\mathbf{I} - \mathbf{H'(HH')}^{-1}\mathbf{H})\mathbf{Z}(\mathbf{I} - \mathbf{G'(GG')}^{-1}\mathbf{G})$$

$$\mathbf{Z'(I} - \mathbf{H'(HH')}^{-1}\mathbf{H})]\} / [(b-1)(t-1) - (b-r-1)(t-q-1)] \quad (3.5.2)$$

For the example in this section,

$$\operatorname{tr}(\mathbf{Z'Z}) = 258,651.73$$

and

$$\operatorname{tr}[(\mathbf{I} - \mathbf{H'(HH')}^{-1}\mathbf{H})\mathbf{Z}(\mathbf{I} - \mathbf{G'(GG')}^{-1}\mathbf{G})$$

$$\cdot \mathbf{Z'(I} - \mathbf{H'(HH')}^{-1}\mathbf{H})] = 254,539$$

Thus,

$$\hat{\sigma}_p^2 = (258,651.73 - 254,539)/(8 - 3) = 822.55$$

with 5 degrees of freedom. This estimator is the same as that previously obtained by deleting the observations that had been judged to be the cause of the nonadditivity in the data. However, this need not always be the case. To complete the analysis of the data in Table 3.3, one will need LSD values for comparing some of the cell values. The 5% LSD for comparing cells is

$$\text{LSD}_{.05} = t_{.025,5} \cdot \hat{\sigma}_p \sqrt{2}$$

$$= 2.571 \cdot 28.68 \cdot 1.414$$

$$= 104.3$$

There was no interaction between nitrogen levels 2 and 3 and the phosphorus levels. A 5% LSD value for comparing these two nitrogen levels after averaging across the five levels of phosphorus is

$$\text{LSD}_{.05} = t_{.025,5} \cdot \hat{\sigma}_p \sqrt{2/5} = 2.571 \cdot 28.68 \cdot .632 = 46.6$$

There was also no interaction between phosphorus levels 3 and 4 and nitrogen levels; the 5% LSD value for comparing phosphorus levels 3 and 4 after averaging across the three levels of nitrogen is

$$\text{LSD}_{.05} = t_{.025,5} \cdot \hat{\sigma}_p \sqrt{2/3} = 2.571 \cdot 28.68 \cdot .816 = 60.2$$

Finally, there was also no interaction between phosphorus levels and the two higher levels of nitrogen. To determine the effect of phosphorus when used with either of the higher levels of nitrogen, it is advisable to average across the two higher levels of nitrogen. The 5% LSD value for these comparisons is

$$\text{LSD}_{.05} = t_{.025,5} \cdot \hat{\sigma}_p \sqrt{2/2} = 2.571 \cdot 28.68 \cdot 1 = 73.7$$

# 4

# Half-Normal Plots

## CHAPTER OUTLINE

Half-normal plotting is a technique introduced by Daniel (1959) to help analyze nonreplicated $2^n$ factorial experiments and fractional replications of such experiments. The plotting technique can be used to estimate the experimental error in nonreplicated experiments as well as to make judgments about which effects are likely to be significant. The technique will be introduced in this chapter and used throughout many of the remaining chapters of this book.

Half-normal plots are very easy to construct if one uses an appropriate computing package. All one needs to construct nice plots is software that will (1) rank or order observed effects, and (2) compute the inverse of the standard normal probability distribution.

**4.1 CONSTRUCTION OF HALF-NORMAL PLOTS**

**Table 4.1   Data Simulated from an $N(0,100)$ Distribution**

| OBS | X | ABSX | R | V |
|-----|-------|-------|-----|-------|
| 1 | 2.00 | 2.00 | 8 | 0.319 |
| 2 | −6.45 | 6.45 | 20 | 0.935 |
| 3 | 8.96 | 8.96 | 23 | 1.150 |
| 4 | 7.54 | 7.54 | 22 | 1.073 |
| 5 | 14.89 | 14.89 | 28 | 1.732 |
| 6 | 6.49 | 6.49 | 21 | 1.001 |
| 7 | 5.99 | 5.99 | 17 | 0.755 |
| 8 | 17.48 | 17.48 | 29 | 1.960 |
| 9 | −4.35 | 4.35 | 15 | 0.648 |
| 10 | 6.27 | 6.27 | 19 | 0.872 |
| 11 | 1.58 | 1.58 | 7 | 0.275 |
| 12 | 3.37 | 3.37 | 13 | 0.549 |
| 13 | 4.17 | 4.17 | 14 | 0.598 |
| 14 | 13.50 | 13.50 | 27 | 1.569 |
| 15 | 9.36 | 9.36 | 24 | 1.235 |
| 16 | −0.76 | 0.76 | 5 | 0.189 |
| 17 | 0.38 | 0.38 | 3 | 0.105 |
| 18 | −5.61 | 5.61 | 16 | 0.701 |
| 19 | 3.37 | 3.37 | 12 | 0.501 |
| 20 | −0.16 | 0.16 | 1 | 0.021 |
| 21 | 2.90 | 2.90 | 9 | 0.363 |
| 22 | −3.03 | 3.03 | 11 | 0.454 |
| 23 | 0.70 | 0.70 | 4 | 0.147 |
| 24 | 3.00 | 3.00 | 10 | 0.408 |
| 25 | 13.01 | 13.01 | 26 | 1.440 |
| 26 | 0.21 | 0.21 | 2 | 0.063 |
| 27 | 6.16 | 6.16 | 18 | 0.812 |
| 28 | 26.78 | 26.78 | 30 | 2.394 |
| 29 | −0.93 | 0.93 | 6 | 0.232 |
| 30 | −11.70 | 11.70 | 25 | 1.331 |

Suppose $X_1, X_2, \ldots, X_n \sim$ i.i.d. $N(0, \sigma^2)$. To construct a half-normal plot of these $X_i$'s, one first orders the absolute values of $X_1, X_2, \ldots, X_n$. Then, as Daniel proposed, the empirical cumulative distribution functions of these ordered values are plotted against the values themselves on half-normal probability paper. If $X_1, X_2, \ldots, X_n$ are distributed i.i.d. $N(0, \sigma^2)$, the resulting points should fall fairly close to a straight line.

As Daniel proposed, plotting these pairs of points would require the use of half-normal probability paper. It is much easier to obtain equivalent plots in the following way: Let

$$R_i = \text{the rank of } |X_i| \quad i = 1, 2, \ldots, n$$

$$R_i^* = (R_i - .5)/n \quad\quad i = 1, 2, \ldots, n$$

$$P_i = .5(R_i^* + 1) \quad\quad i = 1, 2, \ldots, n$$

And let $V_i$ be such that

$$\int_{-\infty}^{V_i} \frac{1}{2\pi} \exp\left[-\frac{1}{2}u^2\right] du = P_i, \quad i = 1, 2, \ldots, n$$

Then a half-normal plot is obtained by plotting $V_i$ against $|X_i|$ for $i = 1, 2, \ldots, n$. This is illustrated using SAS® in the next section.

**4.1.1**
**A SAS®**
**Program**

In this section a SAS® program is given which will produce a half-normal plot. The SAS® statements required are

```
DATA;
INPUT ID X;
ABSX = ABS(X);
CARDS;
1 X₁
2 X₂
 . .

 . .

 . .
n Xₙ
PROC RANK OUT = RANKS;
VAR ABSX;
RANKS R;
DATA PLOTDATA;
SET RANKS;
RSTAR = (R-.5)/n;
P = RSTAR/2 + .5;
V = PROBIT(P);
PROC PLOT; DATA = PLOTDATA;
PLOT V*ABSX = '*'/VREF = 1;
```

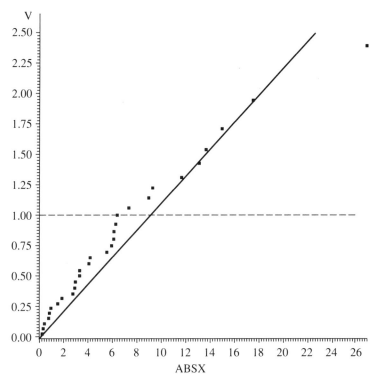

**Figure 4.1   A Half-Normal Plot of the Values in Table 4.1.**

As an example, 32 observations were simulated from a $N(0, 100)$ distribution, i.e., a normal distribution with mean 0 and standard deviation 10. The random deviates obtained are given in Table 4.1 (rounded to two decimal places). **4.1.2 An Example**

A half-normal plot of these data is given in Figure 4.1. This plot was obtained using SASGRAPH® and plotting $V$ against ABSX.

Suppose $X_1, X_2, \ldots, X_n \sim$ i.i.d. $N(0, \sigma^2)$. An approximate estimate of $\sigma$ is given by $\hat{\sigma}$, where $\hat{\sigma}$ is the absolute value of the $X_i$ whose corresponding value of $V_i$ is closest to 1. This is because this value of $X_i$ is approximately one standard deviation from zero. For the data in Section 4.1.2, the sixth observation has a $V_i$ whose value is 1.001, thus an estimate of $\sigma$ is **4.2 ESTIMATING $\sigma$ FROM A HALF- NORMAL PLOT**

$$\hat{\sigma} = |X_6| = 6.49$$

Another method for estimating $\sigma$ is to sketch a line through the origin and the points in Figure 4.1, then take the abscissa of the intersection of this line and the horizontal line through $V = 1$ as the estimate of $\sigma$. For the plot in Figure 4.1, we get $\hat{\sigma} = 8.9$.

**4.3
USE OF
HALF-
NORMAL
PLOTS
IN INTER-
PRETING
NONREPLI-
CATED
EXPER-
IMENTS**

To illustrate how half-normal plots are used to help analyze nonreplicated experiments, consider a $2^5$ factorial experiment which produced the standardized effects given in Table 4.2. The effects are standardized in the sense that var(Effect Estimate) = $\sigma^2$.

A half-normal plot of the effects is in Figure 4.2. If none of the effects is significant, i.e., they are absolute values of contrasts whose expectations are equal to zero, they should fall on a straight line passing through the origin. In Figure 4.2 all of the effects except $B$, $A$, and $D$ appear to fall on a straight line. These three effects would appear to be significant since they do not fall on this straight line.

If all of the effects in Table 4.2 fell on a straight line, $\sigma$ would be estimated by $\hat{\sigma} = 1.688$, the effect whose $V$ value is closest to 1. However, for this example, it is clear that not all of the effects fall on a straight line. When this is the case, one should replot the $V$ values after removing the effects believed to be significant, i.e., the three largest values for this experiment. This plot is shown in Figure 4.3. These points do appear to fall on a straight line, and a revised estimate of $\sigma$ is $\hat{\sigma} = 1.28$, the abscissa of the intersection of the two lines in Figure 4.3.

**Table 4.2  Absolute Values of the Standardized Effects
of a $2^5$ Factorial Experiment**

| Effect | Value | Effect | Value |
|--------|-------|--------|-------|
| $A$ | 5.938 | $E$ | .188 |
| $B$ | 7.000 | $A*E$ | 2.000 |
| $A*B$ | .062 | $B*E$ | .906 |
| $C$ | .281 | $A*B*E$ | .688 |
| $A*C$ | 1.688 | $C*E$ | .219 |
| $B*C$ | .938 | $A*C*E$ | 1.063 |
| $A*B*C$ | .656 | $B*C*E$ | .875 |
| $D$ | 4.781 | $A*B*C*E$ | .438 |
| $A*D$ | 1.656 | $D*E$ | 1.656 |
| $B*D$ | 2.906 | $A*D*E$ | .013 |
| $A*B*D$ | 1.812 | $B*D*E$ | 1.219 |
| $C*D$ | .125 | $A*B*D*E$ | .969 |
| $A*C*D$ | 1.031 | $C*D*E$ | .562 |
| $A*B*C*D$ | 1.469 | $A*C*D*E$ | 1.812 |
| $A*B*C*D$ | 1.469 | $B*C*D*E$ | .500 |
| | | $A*B*C*D*E$ | 2.406 |

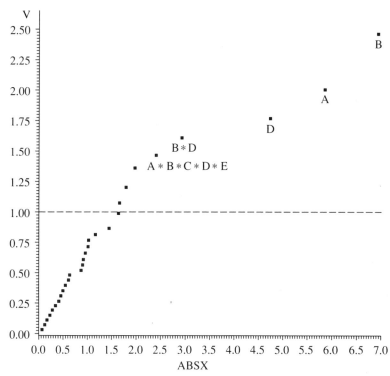

**Figure 4.2   A Half-Normal Plot of the Effects in Table 4.2.**

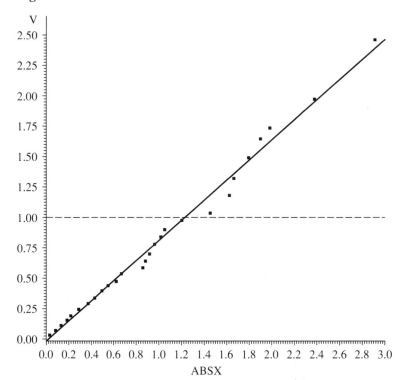

**Figure 4.3   A Half-Normal Plot of the Effects in Table 4.2 After Removing the Three Largest Effects.**

Clearly, in order for the half-normal plot to be most useful, the majority of the effects estimated must correspond to null effects—factors having no real effects. The technique of half-normal plotting is very useful in exploratory situations where one is looking for those factors most likely to have significant effects. These factors could then be studied more accurately in another experiment.

The half-normal plotting method is applied to several other experiments in many of the remaining chapters of this book.

# 5

# Analyzing a Two-Way Treatment Structure in a Split-Plot Design with No Replications

**CHAPTER OUTLINE**

any experiments are conducted where a panel of judges or experts is required to evaluate an experimental unit for some characteristic such as sweetness, saltiness, firmness, or color. Some researchers incorrectly analyze such data as though the scores provided by the different panelists or judges represent independent replications of the product being sampled. However, if the panelists are evaluating the same experimental unit (the quantity to which the treatment has been randomly assigned), then the scores cannot be considered as independent replications. If there is only one independent sampling of each treatment, even though each sample may be measured many times, no conventional statistical analysis method is available for analyzing this type of data. This chapter illustrates, by example, some generalizations of the techniques described in Chapters 1 to 4 which can be used to try to salvage information from such an experiment.

## 5.1
## A SAUSAGE
## EXAMPLE

A food scientist wished to compare 12 different methods for making sausages. For these comparisons, the scientist made one batch of sausage mixture for each of the 12 methods. Each batch was large enough to produce six sausages. Four of these six sausages were selected at random and evaluated independently by four different judges for saltiness. The resulting data are given in Table 5.1.

## 5.2
## AN
## INCORRECT
## ANALYSIS
## OF THE
## SAUSAGE
## DATA

Some researchers have incorrectly analyzed data similar to that in Table 5.1 by assuming that the experimental design for this experiment is a randomized complete block design with judges representing blocks and then taking the Judge * Sausage mean square to be an appropriate estimate of experimental error with which to compare treatment effects. This analysis would be correct if there had been 48 independent batches of sausage made and one sausage being evaluated from each batch. The analysis of variance table for this incorrect analysis is shown in Table 5.2.

### Table 5.1 Sausage Saltiness Evaluations

| | | | | | | SAUSAGE TYPE | | | | | | |
|---|---|---|---|---|---|---|---|---|---|---|---|---|
| Judge | 1 | 2 | 3 | 4 | 5 | 6 | 7 | 8 | 9 | 10 | 11 | 12 |
| A | 2.4 | 2.3 | 2.6 | 6.0 | 4.6 | 3.8 | 4.6 | 3.6 | 0.5 | 2.2 | 4.3 | 9.0 |
| B | 2.5 | 2.0 | 2.5 | 5.5 | 4.1 | 2.3 | 4.5 | 3.1 | 0.3 | 1.8 | 4.5 | 7.0 |
| C | 3.0 | 3.0 | 3.0 | 6.0 | 3.9 | 4.2 | 5.0 | 3.9 | 1.1 | 2.5 | 4.7 | 7.9 |
| D | 2.7 | 2.8 | 2.6 | 5.8 | 4.9 | 4.4 | 4.9 | 4.0 | 1.3 | 2.3 | 4.6 | 8.5 |
| Means | 2.70 | 2.53 | 2.68 | 5.83 | 4.38 | 3.68 | 4.75 | 3.65 | 0.80 | 2.20 | 4.53 | 8.10 |

**Table 5.2   An Incorrect Analysis of the Sausage Data**

| Source of Variation | df | SS | MS | F | P |
|---|---|---|---|---|---|
| Sausage type | 11 | 160.64 | 14.60 | 108.11 | <.01 |
| Judge | 3 | 3.94 | 1.31 | 9.72 | <.01 |
| Error = Sausage * Judge | 33 | 4.46 | 0.135 | | |

The major problem with the above experiment is that there is no measure of the experimental error associated with making sausage batches since each type of sausage was only made once. A correct model for the experiment described in Section 5.1 is

$$y_{ij} = \mu + S_i + \delta_i + J_j + (SJ)_{ij} + \epsilon_{ij}$$
$$i = 1, 2, \ldots, 12, \quad j = 1, 2, 3, 4 \tag{5.2.1}$$

where $S_i$ is the effect due to Sausage Type $i$, $J_j$ is the effect due to Judge $j$, $(SJ)_{ij}$ is the effect due to interactions between Sausage Type and Judge, $\delta_i$ is the error associated with batch making (i.e., the amount of saltiness may vary from batch to batch), and $\epsilon_{ij}$ is the error associated with variations within a batch, measurement errors, etc. It is also assumed that $\delta_i \sim$ i.i.d. $N(0, \sigma_\delta^2)$, $\epsilon_{ij} \sim$ i.i.d. $N(0, \sigma_\epsilon^2)$, and $\delta_{i'}$ are statistically independent of $\epsilon_{ij}$ for all $i'$, $i$, and $j$.

This model is similar to those described [see Milliken and Johnson (1984), Chapters 6 and 24] for split-plot experiments with sausage types representing the whole-plot treatments and judges representing the subplot treatments, except that in the present experiment, the whole-plot treatments have not been replicated.

The expected values of the mean squares for each of the rows in the analysis of variance table in Table 5.2 are given in Table 5.3. These expected values have been calculated in two ways: (1) assuming $(SJ)_{ij}$ is in model (5.2.1) and (2) assuming $(SJ)_{ij}$ is not in the model. In either case, it is clear that the ratio of the Sausage Type mean square to the Error

**Table 5.3   Expected Mean Squares for Table 5.2**

| Source of Variation | Expected Mean Squares* | Expected Mean Squares† |
|---|---|---|
| Sausage type | $\sigma_\epsilon^2 + 4\sigma_\delta^2 + Q_1(S, S*J)$ | $\sigma_\epsilon^2 + 4\sigma_\delta^2 + Q_4(S)$ |
| Judge | $\sigma_\epsilon^2 \qquad\qquad + Q_2(J, S*J)$ | $\sigma_\epsilon^2 \qquad\qquad + Q_5(J)$ |
| Error | $\sigma_\epsilon^2 \qquad\qquad + Q_3(S*J)$ | $\sigma_\epsilon^2$ |

*Assuming $S*J$ interaction.
†Assuming there is no $S*J$ interaction.

mean square is not an appropriate test statistic for comparing sausage types. In either case this ratio could be large if $\sigma_\delta^2$ is large. Unfortunately, there is no way to determine if $\sigma_\delta^2$ is large since there are no independent replications of sausage types, the whole-plot treatments.

## 5.3 TESTING FOR SAUSAGE TYPE * JUDGE INTERAC-TION

Whether one should test for Judge * Sausage Type interaction or not may depend on the kinds of judges one is using. If the judges are chosen on the basis of their expertise (i.e., they are experts), then it is always important to test for Judge * Sausage Type interaction. This is true because the judges should be evaluating the same characteristics of the samples, and they should agree on the relative differences in these characteristics from sample to sample. For example, if the judges are evaluating the samples for saltiness, they should generally agree on which samples are most salty, which samples are least salty, etc. One is generally not interested in whether they score the samples exactly the same or not, but in whether they agree on the relative comparisons between the samples. If the judges cannot agree on the relative comparisons between the samples, they could hardly be called experts. Thus if most of the judges are in agreement with each other except for perhaps one or two judges, then the judges not in agreement with the majority should probably be removed and the data reanalyzed.

If the judges are not chosen for their expertise, i.e., the judges may be a group of randomly chosen consumers, then the test for interaction is generally not as important. In this case the researcher's interest is almost entirely on an average judge rating for each sample.

The techniques of Chapters 1 to 3 can be used to test for Sausage Type * Judge interaction in this kind of an experiment. Unfortunately there is no simple test for comparing sausage types for either the case where there is significant interaction or the case where there is not, as was shown by the expected mean squares in Table 5.3. In an attempt to salvage some information from this experiment, the methods of Chapter 4 are used to try to compare sausage types in the next section. Chapter 4 methods will be applied to independent contrasts of the means of the judge scores for each sausage type. If one or more of the judges is not consistent with the other judges, it may not be appropriate to use that judge's data when calculating sausage type averages. For this reason it is important to determine whether there is significant Judge * Sausage Type interaction. This is done by using the characteristic root test for interaction described in Chapter 1.

To test for interaction between sausage types and judges, the additive model

$$y_{ij} = \mu + S_i^* + J_j + \epsilon_{ij}^* \tag{5.3.1}$$

is first fit to the data. In the above model $S_i^* = S_i + \delta_i$ and $\epsilon_{ij}^* = (SJ)_{ij} + \epsilon_{ij}$. After fitting the model in (5.3.1), the residual matrix $\mathbf{Z}$ is determined. The nonzero characteristic roots of $\mathbf{Z'Z}$ are 2.693, 1.307, and .457.

$U_1 = .604$ in (1.11.1). This is not significant at the .10 level. Generally, in these kinds of experiments, comparisons between judges are not of interest. If such comparisons are of interest and there is no significant Judge $*$ Sausage Type interaction, the value of $F = 9.72$ in Table 5.2 is an appropriate test statistic for comparing judges and the corresponding significance probability is correct. Thus one would conclude that there is a significant difference between judges. If it is important to compare judge means to each other, the appropriate LSD for such pairwise comparisons is given by

$$\text{LSD}_\alpha = t_{\alpha/2,33} \cdot \hat{\sigma}_\epsilon \cdot \sqrt{(2/t)}$$
$$= 2.042 \cdot \sqrt{.135} \cdot \sqrt{2/12}$$
$$= .306$$

for $\alpha = .05$.

In model (5.3.1), $S_i^* = S_i + \delta_i$ and $\epsilon_{ij}^* = (SJ)_{ij} + \epsilon_{ij}$. Without independent replications of the experiment, it is not possible to estimate both $S_i$ and $\delta_i$, nor is it possible to estimate both $(SJ)_{ij}$ and $\epsilon_{ij}$. Because of this we say that $\delta_i$ is confounded with $S_i$ and that $(SJ)_{ij}$ is confounded with $\epsilon_{ij}$. In the next section an attempt is made to determine whether there are significant differences between sausage types. The experimental error for sausage types, $\sigma_\epsilon^2 + 4\sigma_\delta^2$, is also estimated by using the methods of Chapter 4.

To use the methods of Chapter 4 to determine if there are significant differences between sausage types, one must first select a set of orthogonal contrasts of the sausage type means. The set chosen should

**5.4 COMPARING SAUSAGE MEANS**

**Table 5.4   Contrasts of Sausage Type Means**

| | | | | | SAUSAGE TYPE | | | | | | | | |
|---|---|---|---|---|---|---|---|---|---|---|---|---|---|
| Contrast | 8 | 3 | 11 | 4 | 2 | 9 | 5 | 1 | 7 | 12 | 6 | 10 | Estimate |
| 1 | 1 | 0 | −1 | 0 | 0 | 0 | 0 | 0 | 0 | 0 | 0 | 0 | −1.24 |
| 2 | 1 | −2 | 1 | 0 | 0 | 0 | 0 | 0 | 0 | 0 | 0 | 0 | 2.31 |
| 3 | 0 | 0 | 0 | 1 | 0 | −1 | 0 | 0 | 0 | 0 | 0 | 0 | 7.11 |
| 4 | 0 | 0 | 0 | 1 | −2 | 1 | 0 | 0 | 0 | 0 | 0 | 0 | 1.29 |
| 5 | 0 | 0 | 0 | 0 | 0 | 0 | 1 | 0 | −1 | 0 | 0 | 0 | −0.53 |
| 6 | 0 | 0 | 0 | 0 | 0 | 0 | 1 | −2 | 1 | 0 | 0 | 0 | 3.12 |
| 7 | 0 | 0 | 0 | 0 | 0 | 0 | 0 | 0 | 0 | 1 | 0 | −1 | 8.34 |
| 8 | 0 | 0 | 0 | 0 | 0 | 0 | 0 | 0 | 0 | 1 | −2 | 1 | 2.41 |
| 9 | 1 | 1 | 1 | −1 | −1 | −1 | 0 | 0 | 0 | 0 | 0 | 0 | 1.39 |
| 10 | 0 | 0 | 0 | 0 | 0 | 0 | 1 | 1 | 1 | −1 | −1 | −1 | −1.80 |
| 11 | 1 | 1 | 1 | 1 | 1 | 1 | −1 | −1 | −1 | −1 | −1 | −1 | −3.32 |
| Mean | 3.65 | 2.68 | 4.53 | 5.83 | 2.53 | 0.80 | 4.38 | 2.70 | 4.75 | 8.10 | 3.68 | 2.20 | |

be selected independently of the observed values of the sausage type means. That is, one should not let the observed values of the sausage type means influence the selection of the orthogonal contrasts. One possible approach is to select a set of orthogonal contrasts and then randomly assign the sausage type treatments to the columns of the contrast matrix. This approach is illustrated in Table 5.4. A second method, which is preferred whenever possible, is to let the structure of the treatments themselves determine the contrasts. This approach is illustrated by the example in Chapter 6.

Consider the set of orthogonal contrasts given in the rows of Table 5.4 with sausage types randomly assigned to each column.

The quantities under the Estimate column are computed from

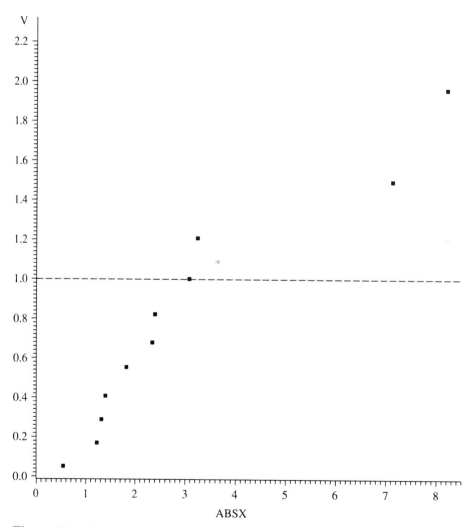

**Figure 5.1   Half-Normal Plot for Sausage Types.**

$$\sqrt{b}\left(\sum c_i \bar{y}^*_{i\cdot}\right)\Big/\sqrt{\sum c_i^2}$$

where $\bar{y}^*_{1\cdot}, \bar{y}^*_{2\cdot}, \ldots, \bar{y}^*_{12\cdot}$ are equal to 3.65, 2.68, ..., 2.20, respectively, and $c_1, c_2, \ldots, c_{12}$ are the contrast coefficients. For example, $-1.24$ in the first row of Table 5.4 is given by $\sqrt{4}(3.65 - 4.53)/\sqrt{1^2 + (-1)^2} = -1.24$. The quantities in the Estimate column can be considered as a random sample from $N(0, \sigma_\epsilon^2 + 4\sigma_\delta^2)$, if there are no differences between sausage types. A half-normal plot of the values in the Estimate column is given in Figure 5.1.

Examination of the plot in Figure 5.1 reveals two contrasts, (7) and (3), which appear to be significantly larger than the rest. Contrast (7) compares Sausage Type 10 to Sausage Type 12, and contrast (3)

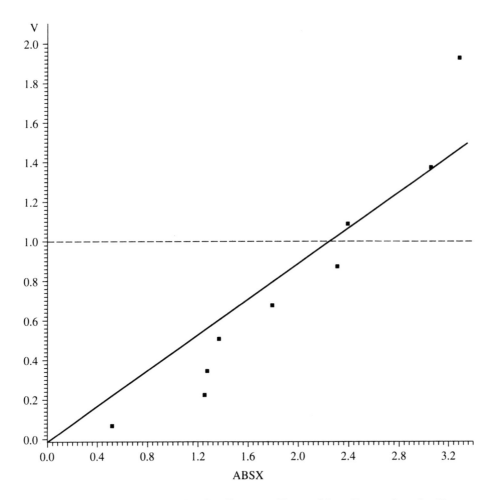

**Figure 5.2   Half-Normal Plot for Sausage Types After Removing the Two Largest Effects.**

**Table 5.5  Pairwise Differences Between Sausage Type Means**

| Sausage | 2 | 3 | 4 | 5 | 6 | 7 | 8 | 9 | 10 | 11 | 12 |
|---------|------|-------|-------|-------|-------|-------|-------|-------|-------|-------|--------|
| 1 | 0.17 | 0.02 | −3.13 | −1.68 | −0.98 | −2.05 | −0.95 | 1.90 | 0.50 | −1.83 | −5.40* |
| 2 | . | −0.15 | −3.30 | −1.85 | −1.15 | −2.22 | −1.12 | 1.73 | 0.33 | −2.00 | −5.57* |
| 3 | . | . | −3.15 | −1.70 | −1.00 | −2.07 | −0.97 | 1.88 | 0.48 | −1.85 | −5.42* |
| 4 | . | . | . | 1.45 | 2.15 | 1.08 | 2.18 | 5.03* | 3.63 | 1.30 | −2.27 |
| 5 | . | . | . | . | 0.70 | −0.37 | 0.73 | 3.58 | 2.18 | −0.15 | −3.72 |
| 6 | . | . | . | . | . | −1.07 | 0.03 | 2.88 | 1.48 | −0.85 | −4.42 |
| 7 | . | . | . | . | . | . | 1.10 | 3.95 | 2.55 | 0.22 | −3.35 |
| 8 | . | . | . | . | . | . | . | 2.85 | 1.45 | −0.88 | −4.45 |
| 9 | . | . | . | . | . | . | . | . | −1.40 | −3.73 | −7.30* |
| 10 | . | . | . | . | . | . | . | . | . | −2.33 | −5.90* |
| 11 | . | . | . | . | . | . | . | . | . | . | −3.57 |

*Significant at approximately the .05 level.

compares Sausage Type 4 with Sausage Type 9. Thus it is reasonable to conclude that Sausage Type 12 is significantly saltier than Sausage Type 10 and that Sausage Type 9 is significantly less salty than Sausage Type 4. An examination of the means in Table 5.1 along with the above information tends to make one believe that Sausage Type 12 is significantly saltier than the rest, that Sausage Type 9 is significantly less salty than the rest, and that the rest are approximately equal in saltiness.

To estimate $(\sigma_\epsilon^2 + 4\sigma_\delta^2)^{1/2}$, we first construct another half-normal plot of the contrasts in Table 5.4 leaving out the largest two contrasts. This plot is shown in Figure 5.2. The abscissa of the intersection of the two lines on this plot is 2.23. Thus

$$(\hat{\sigma}_\epsilon^2 + 4\hat{\sigma}_\delta^2)^{1/2} = 2.23$$

It is interesting to note that $t$-type statistics corresponding to the contrasts (3) and (7) are $7.11/2.23 = 3.19$ and $8.34/2.23 = 3.74$, respectively. These $t$-type statistics, having values greater than 2, give further evidence that contrasts (3) and (7) are significantly different from zero.

Table 5.5 shows the pairwise differences between all pairs of sausage type means. Those which have absolute values greater than

$$2 \cdot (\hat{\sigma}_\epsilon^2 + 4\hat{\sigma}_\delta^2)^{1/2} = 2 \cdot 2.23 = 4.46$$

have been indicated with an asterisk.

Examination of Table 5.5 indicates that Sausage Type 12 is significantly saltier than Sausage Types 1, 2, 3, 9, and 10 and that Sausage Type 9 is significantly less salty than Sausage Types 4 and 12. Given the small number of independent replications in this experiment, the above statements are about all that one may safely say about these sausage types.

# 6

# A Three-Way Treatment Structure with No Replications

## CHAPTER OUTLINE

n this chapter an experiment is considered where the basic treatment structure is three-way and there are no independent replications of these three-way treatment combinations. In order to use the methods of Chapters 1 to 3 on three-way and higher-order treatment structures, one must look at the experiment as though it is two-way. For the three-way this can be done by letting all possible combinations for two of the factors in the three-way represent one of the factors in the two-way and letting the levels of the third factor in the three-way represent the second factor in the two-way. That is, if one had a three-way treatment structure with factors $A$, $B$, and $C$, one could consider the experiment as a two-way with factors $A*B$ and $C$, $A*C$ and $B$, or $A$ and $B*C$.

It can be shown that if there is no two-way interaction between the levels of $A*B$ and $C$, then there can be no three-way interaction between the levels of $A$, $B$, and $C$. However, if there is interaction between the levels of $A*B$ and $C$, there may or may not be a three-way interaction. In this case it would be wise to look for interaction between the levels of $A*C$ and $B$ and/or between the levels of $B*C$ and $A$.

## 6.1
## AN EXAMPLE

A researcher wanted to evaluate the effects that certain factors have on the quality of color photographs. The factors being studied were: (1) the type of chemical used in the developing solution, (2) the length of time the film is in the solution, and (3) the method of finishing. There were three different chemicals, two lengths of time, and two methods of finishing being considered. A roll of film, which provides eight pictures, was shot for each Chemical*Time*Finish combination. Each of the films was exposed under similar and nearly ideal conditions. The eight pictures from each roll were examined by six experts, and each expert scored each picture on a quality scale ranging from 1 to 10, with 10 being perfect. The scores on the eight pictures were totaled for each expert and the results are given in Table 6.1.

Since all of the experts are judging the same eight photographs, their ratings cannot be considered as independent replications of the 12 Chemical*Time*Method combinations. In this experiment, only one roll of film was shot for each Chemical*Time*Method combination, and hence there are no independent replications of these 12 treatment combinations. To compare the effects of Chemicals, Time, and Methods, one should average the experts' scores for each treatment combination and analyze these means. However, before averaging the experts' scores, it would be worth determining whether experts agree or disagree on the relative merits of the photographs. That is, determine whether there is any significant interaction between experts and the Chemical*Time*Method treatment combinations. It does not matter if one or more experts tend to rate photographs higher or lower than the others provided that they are consistently higher or lower across all 12 treatment combinations. If there is significant Expert*Treatment Combinations interaction, then the

**Table 6.1  Photographic Quality Scores**

| | CHEMICAL 1 | | | | CHEMICAL 2 | | | | CHEMICAL 3 | | | |
| | 10 minutes | | 20 minutes | | 10 minutes | | 20 minutes | | 10 minutes | | 20 minutes | |
| | Method | | Method | | Method | | Method | | Method | | Method | |
| Expert | 1 | 2 | 1 | 2 | 1 | 2 | 1 | 2 | 1 | 2 | 1 | 2 |
|---|---|---|---|---|---|---|---|---|---|---|---|---|
| 1 | 40 | 62 | 38 | 60 | 45 | 72 | 47 | 60 | 35 | 56 | 25 | 40 |
| 2 | 35 | 56 | 32 | 50 | 56 | 56 | 53 | 66 | 50 | 48 | 35 | 36 |
| 3 | 45 | 60 | 50 | 48 | 50 | 63 | 46 | 56 | 35 | 52 | 30 | 32 |
| 4 | 42 | 54 | 40 | 61 | 45 | 75 | 52 | 64 | 40 | 45 | 24 | 31 |
| 5 | 48 | 60 | 46 | 55 | 55 | 67 | 56 | 72 | 43 | 59 | 34 | 34 |
| 6 | 45 | 65 | 35 | 53 | 48 | 68 | 47 | 70 | 35 | 52 | 34 | 38 |
| Mean | 42.50 | 59.50 | 40.17 | 54.50 | 49.83 | 66.83 | 50.17 | 64.67 | 39.67 | 52.00 | 30.33 | 35.17 |

methods of Chapter 3 can help determine which experts are consistent with each other and which ones are not. If one or two experts are responsible for the interaction, then those experts could be removed from the analysis and treatment combination means could be recalculated from the remaining experts. If all experts seem to be interacting with treatments, it is probably time to find some new experts or spend some time training or retraining the experts being used. In this case, there is probably not much point in spending any further time in analyzing the data.

**6.2 TESTING FOR EXPERT\*TREATMENT INTERACTION**

To test for Expert\*Treatment interaction, the data in Table 6.1 are analyzed as though they made up a $6 \times 12$ two-way experiment with the 6 experts and the 12 different treatment combinations representing the two factors in the two-way structure. The situation in this case is similar to that observed in Chapter 5, and hence the characteristic root test described in Chapter 1 is an appropriate test for testing for Expert\*Treatment interaction. The characteristic roots of $\mathbf{Z'Z}$, when $\mathbf{Z}$ is the $6 \times 12$ matrix of residuals resulting from fitting an additive model to the data, are 638.694, 371.441, 244.816, 124.317, and 52.954. The residual sum of squares from the additive model is RSS $= 1432.222$. The test statistic for testing for interaction is

$$U_1 = \ell_1/\text{RSS} = 638.694/1432.222 = .446$$

The 10% critical point from Table A.1 is .5001, and since $U_1 < .5001$, this value of $U_1$ is not significant at the .10 level. Also, the value of $U_2 = \ell_2/(\text{RSS} - \ell_1) = .4681$ is not significant. Hence, there appears to be no significant Judge\*Treatment interaction and it is appropriate to analyze the means of the 6 experts' scores for each of the 12 treatment combinations.

**6.3 ANALYZING TREATMENTS FOR INTERACTION**

Upon averaging the experts' scores, one gets the results in Table 6.2. Since two of these factors have only two levels each, these two will need to be combined into one factor in order to use the multiplicative interaction model on these data (remember both rows and columns must have three or more levels in order to use the multiplicative interaction model).

The two nonzero eigenvalues of $\mathbf{Z'Z}$, where $\mathbf{Z}$ is the $3 \times 4$ matrix of residuals after fitting an additive model to the data in Table 6.2, are 118.218 and 1.083, and RSS $= 119.301$.

The characteristic root test for interaction applied to the data in Table 6.2 gives

$$U_1 = \ell_1/\text{RSS}$$
$$= 118.218/119.301$$
$$= .991$$

**Table 6.2  Photograph Quality Composite Scores**

| | TIME | | | | |
| | 10 | | 20 | | |
| | METHOD | | | | |
| Chemical | 1 | 2 | 1 | 2 | Means |
| 1 | 42.5 | 59.5 | 40.167 | 54.5 | 49.167 |
| 2 | 49.833 | 66.833 | 50.167 | 64.667 | 57.875 |
| 3 | 39.667 | 52.0 | 30.333 | 35.167 | 39.292 |
| Means | 44.0 | 59.444 | 40.222 | 51.444 | 48.778 |

Using the critical values in Table A.1, this value of $U_1$ is significant at the .05 level but not at the .01 level since $U_1 > .987$, the 5% critical point, and $U_1 < .998$, the 1% critical point.

The estimate of experimental error based on the multiplicative interaction model is

$$\hat{\sigma}^2 = \ell_2/[(b-1)(t-1) - \nu_1]$$
$$= 1.083/1$$
$$= 1.083$$

The characteristic vectors corresponding to the rows and columns of the data in Table 6.2 are, respectively,

$$\hat{\alpha}' = [-.30 \ -.51 \ .81]$$

and

$$\hat{\gamma}' = [-.59 \ -.25 \ .06 \ .77]$$

In order to determine the cause of the significant interaction, one might next conduct pairwise tests, testing $\alpha_1 = \alpha_2$, $\alpha_1 = \alpha_3$, and $\alpha_2 = \alpha_3$ using the method outlined in Section 3.3. The resulting test statistics are: .286, .015, and .011, respectively. The appropriate 10%, 5%, and 1% critical points from Tables A.5 to A.7, using

$$\delta = \sqrt{\ell_1/\hat{\sigma}^2}$$
$$= \sqrt{118.218/1.083}$$
$$= 10.45$$

are .1914, .0989, and .0205, respectively. Thus $\alpha_1 = \alpha_3$ and $\alpha_2 = \alpha_3$ are rejected at all levels while $\alpha_1 = \alpha_2$ cannot be rejected at any level. This suggests that the third chemical is the cause of the significant interaction in these data.

Assuming the data are additive in the first two rows, the resulting estimate of $\sigma^2$ is $\hat{\sigma}_1^2 = 1.263$, which is based on 3 degrees of freedom. This value of $\hat{\sigma}_1^2$ can also be found from (3.4.1) with $\mathbf{H} = [1 \ -1 \ 0]$. This is also the value of $\hat{\sigma}_p^2$ in (3.4.3), since $\hat{\sigma}_2^2$ in (3.4.2) is equal to zero.

Next consider the columns of the data matrix, and conduct pairwise tests, $\gamma_1 = \gamma_2$, $\gamma_1 = \gamma_3$, $\gamma_1 = \gamma_4$, $\gamma_2 = \gamma_3$, $\gamma_2 = \gamma_4$, $\gamma_3 = \gamma_4$ , using the methods in Section 3.3. The resulting test statistics are .133, .042, .0099, .172, .018, and .035, respectively. The appropriate 10%, 5%, and 1% critical points from Tables A.5 to A.7 are .1827, .0925, and .0183, respectively. All of these hypotheses can be rejected at the 10% level, but those involving column treatment number 4 are the most significant. Thus most of the interaction in these data is due to Row Treatment 3 combined with Column Treatment 4, or, equivalently, most of the interaction is due to the treatment combination, Chemical = 3, Time = 20, and Method = 2. If the data were additive except for the (3,4) cell, the resulting estimate of $\sigma^2$ would be $\hat{\sigma}^2 = 5.409$ with 5 degrees of freedom. This estimator is most easily found by deleting the (3,4) cell from the data set and fitting an additive two-way model to the remaining data. The estimator $\hat{\sigma}_1^2 = 1.263$ will be used for making inferences about several important contrasts of the Chemical∗Temperature∗Method parameter means in Section 6.4.

**6.4
TREATMENT
CONTRASTS**

An orthogonal set of contrasts for making important comparisons among the 12 Chemical∗Temperature∗Method combinations is given in Table 6.3. The parameters $\mu_{ijk}$ represent the result expected for the Chemical $i$, Temperature $j$, and Method $k$ treatment combination.

Values of normalized contrasts for each effect defined in Table 6.3 are given in Table 6.4. The quantity in the Estimate column is computed from

$$\left( \sum_{ijk} c_{ijk} \, \hat{\mu}_{ijk} \right) \Big/ \sqrt{\sum_{ijk} c_{ijk}^2}$$

where the $\hat{\mu}_{ijk}$'s are estimated treatment combination means from the last row of Table 6.3 and the $c_{ijk}$'s are row coefficients given in Table 6.3. For example, the effect defined in the first row of Table 6.3 is computed as

$$(1 \cdot 42.5 + 1 \cdot 59.5 + 1 \cdot 40.167 + 1 \cdot 54.5 + 0 \cdot 49.833$$

$$+ 0 \cdot 66.833 + 0 \cdot 50.167 + 0 \cdot 64.667 - 1 \cdot 39.667 - 1$$

$$\cdot 52.0 - 1 \cdot 30.333 - 1 \cdot 35.167) / [1^2 + 1^2 + 1^2 + 1^2$$

$$+ 0^2 + 0^2 + 0^2 + 0^2 + (-1)^2 + (-1)^2 + (-1)^2$$

$$+ (-1)^2]^{1/2} = 13.97$$

Those effects in Table 6.4 whose absolute values are larger than

# Table 6.3 Chemical*Temperature*Method Contrasts

|  | TREATMENT COMBINATION | | | | | | | | | | | |
|---|---|---|---|---|---|---|---|---|---|---|---|---|
| Effect | $\mu_{111}$ | $\mu_{112}$ | $\mu_{121}$ | $\mu_{122}$ | $\mu_{211}$ | $\mu_{212}$ | $\mu_{221}$ | $\mu_{222}$ | $\mu_{311}$ | $\mu_{312}$ | $\mu_{321}$ | $\mu_{322}$ |
| 1. $C_1$ vs. $C_3$ | 1 | 1 | 1 | 1 | 0 | 0 | 0 | 0 | −1 | −1 | −1 | −1 |
| 2. $C_2$ vs. $(C_1 + C_3)/2$ | 1 | 1 | 1 | 1 | −2 | −2 | −2 | −2 | 1 | 1 | 1 | 1 |
| 3. $T_1$ vs. $T_2$ | 1 | 1 | −1 | −1 | 1 | 1 | −1 | −1 | 1 | 1 | −1 | −1 |
| 4. $M_1$ vs. $M_2$ | −1 | 1 | −1 | 1 | −1 | 1 | −1 | 1 | −1 | 1 | −1 | 1 |
| 5. (1) * (3) | 1 | 1 | −1 | −1 | 0 | 0 | 0 | 0 | −1 | −1 | 1 | 1 |
| 6. (2) * (3) | 1 | 1 | −1 | −1 | −2 | −2 | 2 | 2 | 1 | 1 | −1 | −1 |
| 7. (1) * (4) | −1 | 1 | −1 | 1 | 0 | 0 | 0 | 0 | 1 | −1 | 1 | −1 |
| 8. (2) * (4) | −1 | 1 | −1 | 1 | 2 | −2 | 2 | −2 | −1 | 1 | −1 | 1 |
| 9. (3) * (4) | −1 | 1 | 1 | −1 | −1 | 1 | 1 | −1 | −1 | 1 | 1 | −1 |
| 10. (1) * (3) * (4) | −1 | 1 | 1 | −1 | 0 | 0 | 0 | 0 | 1 | −1 | −1 | 1 |
| 11. (2) * (3) * (4) | −1 | 1 | 1 | −1 | 2 | −2 | −2 | 2 | −1 | 1 | 1 | −1 |
| Treatment Means | 42.5 | 59.5 | 40.167 | 54.5 | 49.833 | 66.833 | 50.167 | 64.667 | 39.667 | 52.0 | 30.333 | 35.1 |

**Table 6.4   Estimates of the Normalized Contrasts**

| Effect | Estimate |
|---|---|
| 1. $C_1$ vs. $C_3$ | 13.97* |
| 2. $C_2$ vs. $(C_1 + C_3)/2$ | $-22.28$* |
| 3. $T_1$ vs. $T_2$ | 10.20* |
| 4. $M_1$ vs. $M_2$ | $-23.09$* |
| 5. $(1) * (3)$ | 6.66* |
| 6. $(2) * (3)$ | $-6.09$* |
| 7. $(1) * (4)$ | $-5.01$* |
| 8. $(2) * (4)$ | 2.96 |
| 9. $(3) * (4)$ | $-3.66$ |
| 10. $(1) * (3) * (4)$ | $-1.71$ |
| 11. $(2) * (3) * (4)$ | 1.05 |

*Indicates approximate significance at the .05 level.

$$t_{.025,3} \cdot \hat{\sigma}_1 = 3.182 \cdot \sqrt{1.263}$$
$$= 3.58$$

have been so indicated with an asterisk.

A half-normal plot of the quantities in the Estimate column of Table 6.4 is shown in Figure 6.1. The half-normal plotting techniques for these data do not appear to give satisfactory results. None of the effects appear to be significant from the half-normal plots. Using the half-normal plotting method of Chapter 4, the estimate of $\sigma$ is 10.20 compared with $\hat{\sigma}_1 = 1.124$ using the methods of the first three chapters. The half-normal methods will generally only be effective when a majority of the calculated effect estimates are estimating error and not real effects in the treatments. In this experiment too many of the effects appear to be measuring significant treatment effects, which renders the half-normal methods ineffective.

The remainder of this chapter lists all of the SAS® statements that were used to compute all of the statistics used in this chapter. This portion of the book can be skipped by non-SAS® users.

```
*  THESE ARE THE SAS STATEMENTS WHICH WERE USED TO PROVIDE
   THE STATISTICS NEEDED FOR THE ANALYSES DISCUSSED IN CHAPTER 6.;

DATA RATING:
  **THESE STATEMENTS INPUT THE DATA IN TABLE 6.1 .;
INPUT CHEM     1-2 TIME     3-4 METHOD     5-6 R1-R6;
SCORE = MEAN(OF R1-R6); *THIS COMPUTES THE MEANS ACROSS EXPERTS;
CTM = 100*CHEM+10*TIME+METHOD; *THIS DEFINES A TRT COMBINATION VARIABLE
                               WHICH IS NEEDED FOR THE ANALYSIS IN
                               SECTION 6.2.;
TM = 10*TIME+METHOD; *THIS DEFINES A TRT COMBINATION VARIABLE WHICH
                     IS NEEDED FOR THE ANALYSIS IN SECTION 6.3.;
```

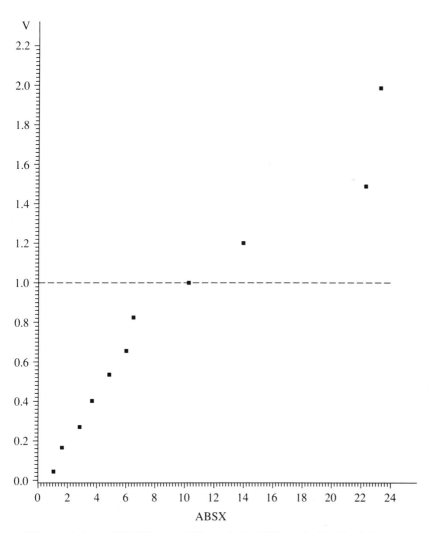

**Figure 6.1   A Half-Normal Plot of the Effects in Table 6.4.**

```
CARDS;
1  1  1  40  35  45  42  48  45
1  1  2  62  56  60  54  60  65
1  2  1  38  32  50  40  46  35
1  2  2  60  50  48  61  55  53
2  1  1  45  56  50  45  55  48
2  1  2  72  56  63  75  67  68
2  2  1  47  53  46  52  56  47
2  2  2  60  66  56  64  72  70
3  1  1  35  50  35  40  43  35
3  1  2  56  48  52  45  59  52
3  2  1  25  35  30  24  34  34
3  2  2  40  36  32  31  34  38
```

```
PROC PRINT;
  TITLE 'THE DATA IN TABLE 6.1';

DATA ONE; SET RATING; * THESE STATEMENTS DEFINE THE VARIABLES REQUIRED FOR
                        THE ANALYSIS IN SECTION 6.2.;
DROP CHEM TIME METHOD TM R1-R6 SCORE;
EXPERT  =  1;  RATING  =  R1;  OUTPUT;
EXPERT  =  2;  RATING  =  R2;  OUTPUT;
EXPERT  =  3;  RATING  =  R3;  OUTPUT;
EXPERT  =  4;  RATING  =  R4;  OUTPUT;
EXPERT  =  5;  RATING  =  R5;  OUTPUT;
EXPERT  =  6;  RATING  =  R6;  OUTPUT;

* THE FOLLOWING 4 STATEMENTS ARE USED TO FIT AN ADDITIVE MODEL TO THE DATA
  AND TO OUTPUT THE RESIDUALS FROM THIS ADDITIVE MODEL TO ANOTHER DATA SET;

PROC GLM;
CLASSES CTM EXPERT;
MODEL RATING  =  CTM EXPERT/SOLUTION;
OUTPUT OUT=RESID RESIDUAL  =  RS;

PROC PRINT DATA=RESID;
  TITLE 'THE DATA AND RESIDUALS FROM AN ADDITIVE MODEL FOR TABLE 6.1.';
  * THE MATRIX PROCEDURE IS USED TO COMPUTE THE STATISTICS REQUIRED FOR ANALYZING
    FOR INTERACTION AND FINDING OUT WHAT CAUSES THE INTERACTION.;

PROC MATRIX FUZZ;
NO_ROW  =  12;  NO_COL  =  6;
FETCH RR DATA=RESID;
NOTE DATA AND RESIDUALS; PRINT RR;
R=RR(*,4);
Z=SHAPE(R,NO_COL);
NOTE PAGE * * * * *  RESIDUAL MATRIX Z * * * * * *;
PRINT Z;
ZPZ=Z`*Z;  ZZP=Z*Z`;
EIGEN ROOTS EZPZ ZPZ; * COMPUTES THE C. ROOTS AND C. VECTORS OF Z`Z;
ROOTS=ROOTS`;
NOTE PAGE EIGENVALUES OF Z``Z;
PRINT ROOTS;
NOTE SKIP  =  2 EIGENVECTORS OF Z``Z;
PRINT EZPZ;
EIGEN RTS EZZP ZZP; * COMPUTES THE C. ROOTS AND C. VECTORS OF ZZ`;
NOTE SKIP  =  2 EIGENVECTORS OF ZZ``;
PRINT EZZP;
LAMDA=EZZP(*,1)`*Z*EZPZ(*,1); * COMPUTES THE VALUE OF LAMDA1HAT.;
NOTE SKIP=2 LAMDA;
PRINT LAMDA;
L1=ROOTS(1,1);
RSS=TRACE(ZPZ);
  U1=L1#/RSS; * COMPUTES THE TEST STATISTIC DEFINED IN SECTION 1.11.4;
NOTE SKIP=2 CHARACTERISTIC ROOT TEST STATISTIC FOR HO:LAMDA =  0;
PRINT U1;
```

```
LAMDA2 = EZZP(* ,2)'* Z* EZPZ(* ,2);   *  COMPUTES  THE  COEF.  OF  A  SECOND  INTERACTION
                                          TERM  FOR  THE  MODEL  IN  (2.1.1);
NOTE  SKIP = 2  LAMDA2;
PRINT  LAMDA2;
NOTE  SKIP  =  2  CHARACTERISTIC  ROOT  TEST  STATISTIC  FOR  HO:LAMDA2 =  0;
L2  =  ROOTS(1,2);
U2 = L2#/(RSS-L1);  *  COMPUTES  THE  TEST  STATISTIC  FOR  HO2  IN  SECTION  2.2.;
PRINT  U2;

* THE  FOLLOWING  TWO  STATEMENTS  PRODUCE  THE  DATA  IN  TABLE  6.2.;
DATA  MEANS;  SET  RATING;
DROP  CTM  TIME  METHOD  R1-R6;

PROC  PRINT;
  TITLE  'THE  DATA  IN  TABLE  6.2';

PROC  GLM;                              * THESE  STATEMENTS  FIT  AN  ADDITIVE  MODEL  TO;
CLASSES  CHEM  TM  ;                    * THE  DATA  IN  TABLE  6.2  AND  OUTPUTS  THE;
MODEL  SCORE  =  CHEM  TM  /SOLUTION;   * RESIDUALS  TO  ANOTHER  DATA  SET;
OUTPUT  OUT = RESID  RESIDUAL  =  RS;

PROC  PRINT  DATA = RESID;
  TITLE  'THE  DATA  AND  RESIDUALS  FROM  AN  ADDITIVE  MODEL  FOR  THE  DATA  IN  TABLE  6.2.';
* THE  MATRIX  PROCEDURE  GIVEN  BELOW  COMPUTES  THE  STATISTICS  NEEDED  FOR  THE
  ANALYSIS  IN  SECTION  6.3.;

PROC  MATRIX  FUZZ;
NO_ROW  =  3;  NO_COL  =  4;
FETCH  RR  DATA = RESID;
NOTE  DATA  AND  RESIDUALS;  PRINT  RR;
R = RR(* ,4);
Z = SHAPE(R,NO_COL);
NOTE  PAGE  *****  RESIDUAL  MATRIX  Z  ******;
PRINT  Z;
ZPZ = Z'* Z;  ZZP = Z* Z';
EIGEN  ROOTS  EZPZ  ZPZ;  * COMPUTES  THE  C.  ROOTS  AND  C.  VECTORS  OF  Z'Z;
ROOTS = ROOTS';
NOTE  PAGE  EIGENVALUES  OF  Z''Z;
PRINT  ROOTS;
NOTE  SKIP  =  2  EIGENVECTORS  OF  Z''Z;
PRINT  EZPZ;
EIGEN  RTS  EZZP  ZZP;  *  COMPUTES  THE  C.  ROOTS  AND  C.  VECTORS  OF  ZZ';
NOTE  SKIP = 2  EIGENVECTORS  OF  ZZ'';
PRINT  EZZP;
LAMDA = EZZP·* ,1)'* Z* EZPZ(* ,1);  *  COMPUTES  LAMDA1HAT;
NOTE  SKIP = 2  LAMDA;
PRINT  LAMDA;
L1 = ROOTS(1,1);
RSS = TRACE(ZPZ);
  U1 = L1#/RSS;  *COMPUTES  THE  TEST  FOR  INTERACTION  DESCRIBED  IN  SECTION  1.11.4.;
NOTE  SKIP = 2  CHARACTERISTIC  ROOT  TEST  STATISTIC  FOR  HO:LAMDA = 0;
PRINT  U1;
```

```
LAMDA2 = EZZP(* ,2)'* Z* EZPZ(* ,2);
NOTE SKIP = 2 LAMDA2;
PRINT LAMDA2;
NOTE SKIP = 2 CHARACTERISTIC ROOT TEST STATISTIC FOR HO:LAMDA2 = 0;
L2 = ROOTS(1,2);
U2 = L2#/(RSS-L1);
PRINT U2;
H = 1 -1 0;
III = 1;
NOTE SKIP = 2 TEST FOR INTERACTION IN ROWS 1 AND 2;
START: W = Z'* (I(3) - H'* INV(H* H')* H)* Z;      * THESE STATEMENTS COMPUTE THE TEST;
  ROOTS = EIGVAL(W): L1STAR = ROOTS(1,1);          * STATISTICS NEEDED TO DETERMINE      ;
  TSTSTAT =        (RSS - L1)#/(RSS - L1STAR);      * WHICH ROWS CONTRIBUTE TO THE        ;
PRINT H ROOTS L1STAR TSTSTAT;                       * THE INTERACTION.                    ;
III = III + 1;
IF III = 2 THEN GO TO TWO;
IF III = 3 THEN GO TO THREE;
IF III = 4 THEN GO TO FOUR;
TWO: H = 1 0 -1;
  NOTE SKIP = 2 TEST FOR INTERACTION IN ROWS 1 AND 3;
  GO TO START;
THREE: H = 0 1 -1;
  NOTE SKIP = 2 TEST FOR INTERACTION IN ROWS 2 AND 3;
  GO TO START;
FOUR: G = 1 -1 0 0;
NOTE SKIP = 2 TEST FOR INTERACTION IN COLS 1 AND 2;
START2: W = Z* (I(4) - G'* INV(G* G')* G)* Z';      * THESE STATEMENTS ARE USED TO;
  ROOTS = EIGVAL(W); L1STAR = ROOTS(1,1);           * DETERMINE WHICH COLUMNS      ;
  TSTSTAT = (RSS - L1)#/(RSS - L1STAR);             * CONTRIBUTE TO THE            ;
PRINT G ROOTS L1STAR TSTSTAT;                        * INTERACTION.                 ;
III = III + 1;
IF III = 5 THEN GO TO FIVE;
IF III = 6 THEN GO TO SIX;
IF III = 7 THEN GO TO SEVEN;
IF III = 8 THEN GO TO EIGHT;
IF III = 9 THEN GO TO NINE;
IF III = 10 THEN STOP;
FIVE: G = 1 0 -1 0;
  NOTE SKIP = 2 TEST FOR INTERACTION IN COLS 1 AND 3;
  GO TO START2;
SIX: G = 1 0 0 -1 ;
  NOTE SKIP = 2 TEST FOR INTERACTION IN COLS 1 AND 4;
  GO TO START2;
SEVEN: G = 0 1 -1 0;
  NOTE SKIP = 2 TEST FOR INTERACTION IN COLS 2 AND 3;
  GO TO START2;
EIGHT: G = 0 1 0 -1 ;
  NOTE SKIP = 2 TEST FOR INTERACTION IN COLS 2 AND 4;
  GO TO START2;
NINE: G = 0 0 1 -1;
  NOTE SKIP = 2 TEST FOR INTERACTION IN COLS 3 AND 4;
```

```
    GO  TO  START2;
TEN:  STOP;
```

\* THE FOLLOWING 3 STATEMENTS FIT AN ADDITIVE MODEL TO THE FIRST TWO ROWS OF THE DATA IN TABLE 6.2. THIS GIVES THE VARIANCE ESTIMATE OF 1.263.;

```
DATA  ROWS12;  SET  MEANS;  IF  CHEM = 1  OR  CHEM = 2;
PROC  ANOVA;  CLASSES  CHEM  TM;
MODEL  SCORE = CHEM  TM;
```

\* THE FOLLOWING 3 STATEMENTS FIT AN ADDITIVE MODEL TO ALL OF THE DATA EXCEPT THE (3,4) CELL. THIS GIVES THE VARIANCE ESTIMATE OF 5.409.;

```
DATA  ALLBUT34;  SET  MEANS;  IF  CHEM = 3  AND  TM = 22  THEN  DELETE;
PROC  GLM ;  CLASSES  CHEM  TM;
MODEL  SCORE = CHEM  TM;
```

\* THE FOLLOWING GLM PROCEDURE USES ESTIMATE OPTIONS TO COMPUTE EACH OF THE ESTIMATES IN TABLE 6.4.;

```
PROC  GLM  DATA = RATING;
CLASSES  CHEM  TIME  METHOD;
MODEL  SCORE  =  CHEM[TIME[METHOD;
```

\* NOTE: THE DIVISOR FOR AN ESTIMATE IS THE SQRT(C'C/R) WHERE R = THE NUMBER OF LEVELS OF THE OTHER FACTORS THAT ARE BEING AVERAGED OVER.;

```
ESTIMATE  '1'  CHEM   −1  0  1/DIVISOR  =  0.7071068;
ESTIMATE  '2'  CHEM   1   −2  1/DIVISOR  =  1.2247449;
ESTIMATE  '3'  TIME  1   −1  /DIVISOR  =  0.5773503;
ESTIMATE  '4'  METHOD  1   −1  /DIVISOR  =  0.5773503;
ESTIMATE  '5'  CHEM* TIME  1   −1  0  0   −1  1/DIVISOR  =  1.4142136;
ESTIMATE  '6'  CHEM* TIME   −1  1  2   −2   −1  1/DIVISOR  =  2.4494897;
ESTIMATE  '7'  CHEM* METHOD  1   −1  0  0   −1  1/DIVISOR  =  1.4142136;
ESTIMATE  '8'  CHEM* METHOD   −1  1  2   −2   −1  1/DIVISOR  =  2.4494897;
ESTIMATE  '9'  TIME* METHOD  1   −1   −1  1/DIVISOR  =  1.1547005;
ESTIMATE  '10'  CHEM* TIME* METHOD   −1  1  1   −1  0  0  0  0  1   −1   −1  1/DIVISOR =  2.8284271;
ESTIMATE  '11'  CHEM* TIME* METHOD   −1  1  1   −1  2   −2   −2  2   −1  1  1   −1/DIVISOR =  4.8989795;
```

# 7

# $2^n$ Factorial Treatment Structures

## CHAPTER OUTLINE

This chapter considers the analysis of higher-order treatment structures where each of the treatment factors occurs at only two levels. For example a $2^3$ treatment structure experiment would be an experiment involving three treatment factors with each factor at two levels, i.e., a $2 \times 2 \times 2$ experiment. These kinds of experiments are very important for exploratory purposes. They allow researchers to consider experiments which involve large numbers of treatment factors all in one experiment. This is very important because it can be shown that if a researcher wishes to optimize a process (i.e., find that combination of the treatment factors which will give the most desirable outcome), it is almost always more efficient to study all of the factors which are thought to have an influence on the process simultaneously rather than studying each of the factors one-at-a-time by holding all of the other factors fixed. These kinds of factorial experiments are most useful for an initial exploratory study of the process under study. They allow researchers to determine those factors which have the greatest impact on the process under study, and hence experimenters learn which factors should be concentrated on in order to be successful in improving the process. After discovering those factors which have the greatest impact on the process, follow-up experiments can be performed, which should help experimenters fine-tune the process.

Factorial experiments in which all factors occur at only two levels are relatively simple to analyze by hand, at least if one is not measuring very many characteristics of the process. There are many excellent books which illustrate the analysis of such experiments by hand; see for example, Box, Hunter, and Hunter (1978), Cochran and Cox (1957), and Hicks (1982). The emphasis in this book is upon analyzing such experiments with existing statistical computing packages.

**7.1 NOTATION**

The notation that is used in this book for $2^n$ factorial experiments is similar to that used by Hicks (1982) and is illustrated in this section by using a $2^3$ experiment involving treatment factors $A$, $B$, and $C$. For convenience, one of the levels of each factor is called the low level, and the other is called the high level. If a factor is quantitative in nature, temperature for example, then the labeling of a low level and a high level is obvious. However, if a factor is qualitative in nature, for example water-cooled versus air-cooled, such a labeling is not obvious. In this case, it does not matter which level is called the low level and which is called the high level as long as one consistently uses this designation throughout the discussion and analysis of the experiment under consideration.

Table 7.1 shows two ways of designating the responses expected to each of the eight possible treatment combinations in a $2^3$ experiment.

The expressions on the right-hand side of the equal sign in Table 7.1 are only symbolic and are used merely as a technique for simplifying the discussion of $2^n$ factorial experiments. For example, the response

### Table 7.1   Notations for the $2^3$ Factorial Experiment

| A | B | C | Expected Response |
|---|---|---|---|
| Low | Low | Low | $\mu_{000} = (1)$ |
| High | Low | Low | $\mu_{100} = a$ |
| Low | High | Low | $\mu_{010} = b$ |
| High | High | Low | $\mu_{110} = ab$ |
| Low | Low | High | $\mu_{001} = c$ |
| High | Low | High | $\mu_{101} = ac$ |
| Low | High | High | $\mu_{011} = bc$ |
| High | High | High | $\mu_{111} = abc$ |

expected when one uses factor $A$ at a high level, factor $B$ at a low level, and factor $C$ at a high level is defined to be $\mu_{101}$ and is denoted by $ac$. The symbol $ac$ is also sometimes used to represent the treatment combination itself. It is usually clear from the context of the use of these symbols as to whether they refer to the treatment combination under study or to the result expected from this treatment combination. The best estimates of the expected responses to treatment combinations $(1)$ $a$, $b$, $ab$, $c$, $ac$, $bc$, and $abc$ are denoted by $(\hat{1})$ $\hat{a}$, $\hat{b}$, $\widehat{ab}$, $\hat{c}$, $\widehat{ac}$, $\widehat{bc}$, and $\widehat{abc}$, respectively.

A means model for a $2^3$ experiment conducted in a completely randomized design is

$$y_{ijk} = \mu_{ijk} + \epsilon_{ijk} \quad i = 0,1, \quad j = 0,1, \quad \text{and} \quad k = 0,1 \quad (7.1.1)$$

and an effects model for a $2^3$ experiment is given by

$$y_{ijk} = \mu + A_i + B_j + (AB)_{ij} + C_k$$
$$+ (AC)_{ik} + (BC)_{jk} + (ABC)_{ijk} + \epsilon_{ijk}$$
$$i = 0,1, \quad j = 0,1, \quad k = 0,1 \quad (7.1.2)$$

**7.2
MAIN
EFFECTS
AND INTER-
ACTION
EFFECTS**

When using $2^n$ factorial experiments for exploratory purposes, it is important to determine which factors have significant impacts on the process characteristics. This is accomplished by estimating the main effects and interaction effects which are associated with the factorial treatment structure. These effects are defined in this section.

The main effects and interaction effects in $2^n$ factorial experiments can be defined in terms of special contrasts of the expected responses to the possible treatment combinations. For example, in a $2^3$ experiment the main effect of factor $A$, denoted by $A$, is defined by

$$A = \left[-(1) + a - b + ab - c + ac - bc + abc\right]/2^{3-1}$$

In general, the main effect for factor $A$ in a $2^n$ experiment is defined by that contrast of the treatment combinations which has plus signs on all of the treatment combinations which have factor $A$ at a high level and minus signs on all of the treatment combinations which have factor $A$ at a low level divided by $2^{n-1}$. Similarly, the main effect for factor $B$ is defined by that contrast of the treatment combinations which has plus signs on all treatment combinations which have factor $B$ at a high level and minus signs on all treatment combinations which have factor $B$ at a low level divided by $2^{n-1}$. Thus for the $2^3$ experiment the main effect of $B$ is

$$B = \left[-(1) - a + b + ab - c - ac + bc + abc\right]/2^{3-1}$$

The $A*B$ interaction effect is defined by the difference between the effect of $A$ at the high level of factor $B$ and the effect of $A$ at the low level of factor $B$ divided by $2^{n-2}$. For the $2^3$ experiment the effect of $A$ at the high level of factor $B$ is $\left[-b + ab - bc + abc\right]/2$ and the effect of $A$ at the low level of factor $B$ is $\left[-(1) + a - c + ac\right]/2$. Hence the $A*B$ interaction effect is defined by

$$A*B = \left\{-b + ab - bc + abc\right.$$
$$\left. -\left[-(1) + a - c + ac\right]\right\} /2^{3-1}$$
$$= \left[(1) - a - b + ab + c - ac - bc + abc\right]/2^{3-1}$$

In a similar manner, one can define the $A*C$ interaction effect and the $B*C$ interaction effect. To define the $A*B*C$ interaction effect one can take the difference between the $A*B$ interaction at the high level of factor $C$ and the $A*B$ interaction at the low level of factor $C$. Thus, for a $2^3$ experiment, the $A*B*C$ interaction effect is defined by

$$A*B*C = \left\{\left[c - ac - bc + abc\right] - \left[(1) - a - b - ab\right]\right\} /2^{3-1}$$
$$= \left[-(1) + a + b - ab + c - ac - bc + abc\right]/2^{3-1}$$

Table 7.2 shows the signs of the coefficients which define each of the effects in a $2^3$ experiment. Note that the signs for the coefficients of any interaction effect can be found by examining the coefficients for the corresponding main effects. For example, the signs for the $A*B$ interaction effect can be found by multiplying the signs for the $A$ and $B$ effects with multiplication defined by

$$+ \cdot + = - \cdot - = + \quad \text{and} \quad + \cdot - = - \cdot + = -$$

Similarly, the signs for the $A*B*C$ interaction effect can be found by multiplying the signs for the $A*B$ effect and the $C$ effect or by multi-

**Table 7.2   Coefficients for the Effects in a $2^3$ Factorial Experiment**

|        | TREATMENT COMBINATION | | | | | | | |
|--------|-----|-----|-----|-----|-----|-----|-----|-----|
| Effect | (1) | a | b | ab | c | ac | bc | abc |
| Total  | + | + | + | + | + | + | + | + |
| A      | − | + | − | + | − | + | − | + |
| B      | − | − | + | + | − | − | + | + |
| A * B  | + | − | − | + | + | − | − | + |
| C      | − | − | − | − | + | + | + | + |
| A * C  | + | − | + | − | − | + | − | + |
| B * C  | + | + | − | − | − | − | + | + |
| A * B * C | − | + | + | − | + | − | − | + |

plying the signs for the $A*C$ effect and the $B$ effect or by multiplying the signs for the $B*C$ effect and the $A$ effect.

Table 7.3 shows the coefficients which are required to define each of the effects in a $2^4$ factorial experiment. These two examples should make it clear as to how one would define all possible effects in a $2^n$ experiment for $n > 4$.

**Table 7.3   Coefficients for the Effects in a $2^4$ Factorial Experiment**

|        | TREATMENT COMBINATION | | | | | | | | | | | | | | | |
|--------|-----|---|---|----|---|----|----|-----|---|----|----|-----|----|-----|-----|------|
| Effect | (1) | a | b | ab | c | ac | bc | abc | d | ad | bd | abd | cd | acd | bcd | abcd |
| Total     | + | + | + | + | + | + | + | + | + | + | + | + | + | + | + | + |
| A         | − | + | − | + | − | + | − | + | − | + | − | + | − | + | − | + |
| B         | − | − | + | + | − | − | + | + | − | − | + | + | − | − | + | + |
| A * B     | + | − | − | + | + | − | − | + | + | − | − | + | + | − | − | + |
| C         | − | − | − | − | + | + | + | + | − | − | − | − | + | + | + | + |
| A * C     | + | − | + | − | − | + | − | + | + | − | + | − | − | + | − | + |
| B * C     | + | + | − | − | − | − | + | + | + | + | − | − | − | − | + | + |
| A * B * C | − | + | + | − | + | − | − | + | − | + | + | − | + | − | − | + |
| D         | − | − | − | − | − | − | − | − | + | + | + | + | + | + | + | + |
| A * D     | + | − | + | − | + | − | + | − | − | + | − | + | − | + | − | + |
| B * D     | + | + | − | − | + | + | − | − | − | − | + | + | − | − | + | + |
| A * B * D | − | + | + | − | − | + | + | − | + | − | − | + | + | − | − | + |
| C * D     | + | + | + | + | − | − | − | − | − | − | − | − | + | + | + | + |
| A * C * D | − | + | − | + | + | − | + | − | + | − | + | − | − | + | − | + |
| B * C * D | − | − | + | + | + | + | − | − | + | + | − | − | − | − | + | + |
| A * B * C * D | + | − | − | + | − | + | + | − | − | + | + | − | + | − | − | + |

The first important task in analyzing $2^n$ factorial experiments is to find a good estimate of the experimental error variance. Two methods have been proposed for nonreplicated $2^n$ experiments. One method is to compute contrasts of the observed data which correspond to each of the $2^n - 1$ main effects and interactions defined for the $2^n$ experiment, standardize these contrasts so that each has a variance equal to $\sigma^2$, and construct a half-normal plot of the absolute values of these standardized contrasts using the techniques discussed in Chapter 4. The methods of Chapter 4 are then used to find an estimator of $\sigma$.

A second method which has been proposed for estimating $\sigma^2$ in $2^n$ factorial experiments is based on the practical experiences of many experimenters. It has generally been observed that higher-order interactions (four- or five-factor interactions and higher-order interactions) are rarely significant, and hence, most likely correspond to contrasts which measure experimental error rather than effects of the sampled treatment combinations. Therefore, those contrasts which correspond to the higher-order interactions are pooled to provide an estimator of the experimental error variance. The easiest way to do this is by analyzing the results of the experiment using a statistical package which does an analysis of variance and using a model which includes only the main effects and lower-order interactions. Most of these programs will then automatically pool the higher-order interaction effects into the error for the model.

Which of these two methods should be used generally depends on the size of the experiment under study. Complete factorial experiments which have fewer than six factors should generally be analyzed by the half-normal plot technique, while those which have six or more factors should be analyzed by pooling the higher-order interactions to estimate the experimental error variance. Experiments which involve fewer than six factors do not have a sufficient number of high-order interaction effects to provide good estimates of $\sigma^2$, so these experiments are analyzed by using the half-normal plot techniques. It might be noted that experiments which have more than six factors sometimes have so many high-order interaction effects that they provide more information for estimating the error variance than is probably necessary. This makes it possible for experimenters to conduct factorial experiments more efficiently by grouping the treatment combinations into subgroups in a special way and then randomly assigning the treatments in each subgroup to a blocked set of experimental units. Such experiments are discussed in Chapter 8. In factorial experiments which involve a large number of factors, it is often possible to estimate all of the low-order interaction effects and the experimental error without conducting a complete replication of the treatment combinations. For example, a complete replication of a $2^8$ experiment would require 256 experimental runs while the number of three-factor and lower-order effects is only 92. Each of these effects, as well as $\sigma^2$, could easily be estimated with one-half of the 256

**7.3**
**ESTIMATING**
$\sigma^2$ **IN** $2^n$ **FAC-**
**TORIAL**
**EXPER-**
**IMENTS**

experimental runs required for a complete replicate provided that one selects the right half to study. Such experimental plans are called fractional replications of factorial experiments and they are discussed in Chapter 9.

**7.4
ANALYZING
$2^n$ FACTO-
RIAL
EXPER-
IMENTS FOR
SIGNIFICANT
TREATMENT
EFFECTS**

The analysis of $2^n$ factorial experiments is illustrated by an example. A $2^5$ factorial experiment was conducted to study the effects of water, mixing time, temperature, and cooking oil on cake quality using two different mixer types. The factors under study and their levels are given in Table 7.4 and the results of the experiment are given in Table 7.5.

The data in this experiment will be analyzed by both of the methods discussed in Section 7.4. The analysis of variance method will be discussed in Section 7.4.1 and the half-normal plot method will be discussed in Section 7.4.2.

**7.4.1
The Analysis of
Variance
Method
Applied to the
Data in Table
7.5**

The analysis of variance method would not generally be recommended for experiments of this size. It is being used here for illustration purposes only. To use the analysis of variance method (i.e., pooling the high-order interaction effects to estimate $\sigma^2$), the data could be analyzed by using any statistical computing package which will analyze $n$-way cross-classified experiments. The data in Table 7.5 were analyzed using the ANOVA procedure in SAS®. The analysis commands used were:

```
PROC  ANOVA;
CLASSES  W  M  T  C  P;
MODEL  QUALITY  =  W  M  W*M  T  W*T  M*T  W*M*T
C  W*C  M*C  W*M*C  T*C  W*T*C  M*T*C
                          P  W*P  M*P  W*M*P  T*P  W*T*P
M*T*P  C*P  W*C*P  M*C*P  T*C*P;
```

### Table 7.4   Definition of the Treatment Structure in a $2^5$ Factorial Experiment to Study Cake Quality

|  | LEVEL | |
| --- | --- | --- |
| *Variable* | *Low* | *High* |
| W: amount of water | 40 | 120 |
| M: mixing time | 3 | 6 |
| T: temperature | 300 | 500 |
| C: cooking oil | 6 | 12 |
| P: mixer type | $P_1$ | $P_2$ |

**Table 7.5  Results from a $2^5$ Factorial Experiment
to Study Cake Quality**

| | TREATMENT COMBINATION | | | | | |
| Run | W | M | T | C | P | Quality |
|---|---|---|---|---|---|---|
| 27 | 0 | 0 | 0 | 0 | 0 | 4.8 |
| 3 | 1 | 0 | 0 | 0 | 0 | 3.9 |
| 11 | 0 | 1 | 0 | 0 | 0 | 5.0 |
| 19 | 1 | 1 | 0 | 0 | 0 | 2.2 |
| 22 | 0 | 0 | 1 | 0 | 0 | 3.9 |
| 15 | 1 | 0 | 1 | 0 | 0 | 4.2 |
| 5 | 0 | 1 | 1 | 0 | 0 | 3.0 |
| 23 | 1 | 1 | 1 | 0 | 0 | 2.2 |
| 21 | 0 | 0 | 0 | 1 | 0 | 5.7 |
| 14 | 1 | 0 | 0 | 1 | 0 | 2.2 |
| 2 | 0 | 1 | 0 | 1 | 0 | 8.4 |
| 12 | 1 | 1 | 0 | 1 | 0 | 8.3 |
| 17 | 0 | 0 | 1 | 1 | 0 | 5.3 |
| 7 | 1 | 0 | 1 | 1 | 0 | 2.3 |
| 29 | 0 | 1 | 1 | 1 | 0 | 8.6 |
| 25 | 1 | 1 | 1 | 1 | 0 | 8.9 |
| 32 | 0 | 0 | 0 | 0 | 1 | 4.2 |
| 1 | 1 | 0 | 0 | 0 | 1 | 5.0 |
| 16 | 0 | 1 | 0 | 0 | 1 | 5.8 |
| 24 | 1 | 1 | 0 | 0 | 1 | 5.2 |
| 8 | 0 | 0 | 1 | 0 | 1 | 4.6 |
| 10 | 1 | 0 | 1 | 0 | 1 | 4.1 |
| 26 | 0 | 1 | 1 | 0 | 1 | 5.4 |
| 4 | 1 | 1 | 1 | 0 | 1 | 5.2 |
| 30 | 0 | 0 | 0 | 1 | 1 | 2.9 |
| 6 | 1 | 0 | 0 | 1 | 1 | 3.0 |
| 9 | 0 | 1 | 0 | 1 | 1 | 6.7 |
| 28 | 1 | 1 | 0 | 1 | 1 | 6.6 |
| 20 | 0 | 0 | 1 | 1 | 1 | 5.0 |
| 18 | 1 | 0 | 1 | 1 | 1 | 2.7 |
| 31 | 0 | 1 | 1 | 1 | 1 | 7.0 |
| 13 | 1 | 1 | 1 | 1 | 1 | 7.1 |

The model statement used above includes terms for all main effects, two-factor interactions, and three-factor interactions. Those effects which are not included in the model statement, i.e., the four-factor interactions and the five-factor interaction for this example, will be automatically pooled in this analysis to provide an estimate of the experimental error variance. The results of this analysis are shown in Table 7.6.

**Table 7.6   An ANOVA for the Data in Table 7.5 After Pooling High-Order Interactions into Error**

*Dependent Variable: Quality*

| Source | DF | Sum of squares | | Mean square | F value |
|---|---|---|---|---|---|
| Model | 25 | 116.92625000 | | 4.67705000 | 12.63 |
| Error | 6 | 2.22250000 | | 0.37041667 | PR > F |
| Corrected Total | 31 | 119.14875000 | | | 0.0023 |

| R-square | C.V. | Root MSE | | Quality mean | |
|---|---|---|---|---|---|
| 0.981347 | 12.2182 | 0.60861865 | | 4.98125000 | |

| Source | DF | ANOVA SS | F value | PR > F | |
|---|---|---|---|---|---|
| W | 1 | 5.44500000 | 14.70 | 0.0086 | |
| M | 1 | 31.60125000 | 85.31 | 0.0001 | |
| W∗M | 1 | 0.72000000 | 1.94 | 0.2127 | |
| T | 1 | 0.00500000 | 0.01 | 0.9113 | |
| W∗T | 1 | 0.03125000 | 0.08 | 0.7812 | |
| M∗T | 1 | 0.04500000 | 0.12 | 0.7393 | |
| W∗M∗T | 1 | 0.78125000 | 2.11 | 0.1966 | |
| C | 1 | 15.12500000 | 40.83 | 0.0007 | |
| W∗C | 1 | 0.45125000 | 1.22 | 0.3120 | |
| M∗C | 1 | 34.44500000 | 92.99 | 0.0001 | |
| W∗M∗C | 1 | 5.28125000 | 14.26 | 0.0092 | |
| T∗C | 1 | 1.36125000 | 3.67 | 0.1037 | |
| W∗T∗C | 1 | 0.40500000 | 1.09 | 0.3360 | |
| M∗T∗C | 1 | 0.06125000 | 0.17 | 0.6984 | |
| P | 1 | 0.08000000 | 0.22 | 0.6585 | |
| W∗P | 1 | 1.90125000 | 5.13 | 0.0641 | |
| M∗P | 1 | 0.32000000 | 0.86 | 0.3885 | |
| W∗M∗P | 1 | 0.21125000 | 0.57 | 0.4787 | |
| T∗P | 1 | 0.45125000 | 1.22 | 0.3120 | |
| W∗T∗P | 1 | 1.62000000 | 4.37 | 0.0815 | |
| M∗T∗P | 1 | 0.01125000 | 0.03 | 0.8674 | |
| C∗P | 1 | 11.28125000 | 30.46 | 0.0015 | |
| W∗C∗P | 1 | 0.00500000 | 0.01 | 0.9113 | |
| M∗C∗P | 1 | 5.28125000 | 14.26 | 0.0092 | |
| T∗C∗P | 1 | 0.00500000 | 0.01 | 0.9113 | |

The estimate of $\sigma^2$ for the data in Table 7.5 from the ANOVA table in Table 7.6 is $\hat{\sigma}^2 = .3704$ and is based on six degrees of freedom and hence the estimate of $\sigma$ is $\hat{\sigma} = .61$. The significant effects are $W$, $M$, $C$, $M∗C$, $W∗M∗C$, $C∗P$, and $M∗C∗P$. Each of these effects is significant at the .01 level of significance. A discussion of the meaning of these significant effects will be given in Section 7.5.

### Table 7.7 An ANOVA for the Data in Table 7.5 Using a Full Effects Model

*Dependent Variable: Quality*

| Source | DF | Sum of squares | | Mean square | F value |
|---|---|---|---|---|---|
| Model | 31 | 119.14875000 | | 3.84350806 | 99999.99 |
| Error | 0 | 0.00000000 | | 0.00000000 | PR > F |
| Corrected | | | | | |
| Total | 31 | 119.14875000 | | | 0.0000 |

| R-square | C.V. | Root MSE | | Quality mean | |
|---|---|---|---|---|---|
| 1.000000 | 0.0000 | 0.00000000 | | 4.98125000 | |

| Source | DF | ANOVA SS | F value | PR > F |
|---|---|---|---|---|
| W | 1 | 5.44500000 | * | * |
| M | 1 | 31.60125000 | * | * |
| W*M | 1 | 0.72000000 | * | * |
| T | 1 | 0.00500000 | * | * |
| W*T | 1 | 0.03125000 | * | * |
| M*T | 1 | 0.04500000 | * | * |
| W*M*T | 1 | 0.78125000 | * | * |
| C | 1 | 15.12500000 | * | * |
| W*C | 1 | 0.45125000 | * | * |
| M*C | 1 | 34.44500000 | * | * |
| W*M*C | 1 | 5.28125000 | * | * |
| T*C | 1 | 1.36125000 | * | * |
| W*T*C | 1 | 0.40500000 | * | * |
| M*T*C | 1 | 0.06125000 | * | * |
| W*M*T*C | 1 | 0.00000000 | * | * |
| P | 1 | 0.08000000 | * | * |
| W*P | 1 | 1.90125000 | * | * |
| M*P | 1 | 0.32000000 | * | * |
| W*M*P | 1 | 0.21125000 | * | * |
| T*P | 1 | 0.45125000 | * | * |
| W*T*P | 1 | 1.62000000 | * | * |
| M*T*P | 1 | 0.01125000 | * | * |
| W*M*T*P | 1 | 0.40500000 | * | * |
| C*P | 1 | 11.28125000 | * | * |
| W*C*P | 1 | 0.00500000 | * | * |
| M*C*P | 1 | 5.28125000 | * | * |
| W*M*C*P | 1 | 1.28000000 | * | * |
| T*C*P | 1 | 0.00500000 | * | * |
| W*T*C*P | 1 | 0.03125000 | * | * |
| M*T*C*P | 1 | 0.40500000 | * | * |
| W*M*T*C*P | 1 | 0.10125000 | * | * |

**7.4.2**
**The Half-**
**Normal**
**Plot Method**
**Applied to the**
**Data in Table**
**7.5**

One of the easiest ways to obtain the information necessary for a half-normal plot of the data in Table 7.5 is to once again use an ANOVA procedure, this time fitting a model which includes all main effects and all possible interactions. Each of the resulting sums of squares which correspond to the effects in the model, when divided by $\sigma^2$, have single-degree-of-freedom chi-square sampling distributions. Thus the square roots of these sums of squares have sampling distributions which are the same as the absolute values of normally distributed variables. Hence the half-normal plotting technique can be applied to the square roots of the single-degree-of-freedom sums of squares for each of the $2^n - 1$ effects in a $2^n$ experiment.

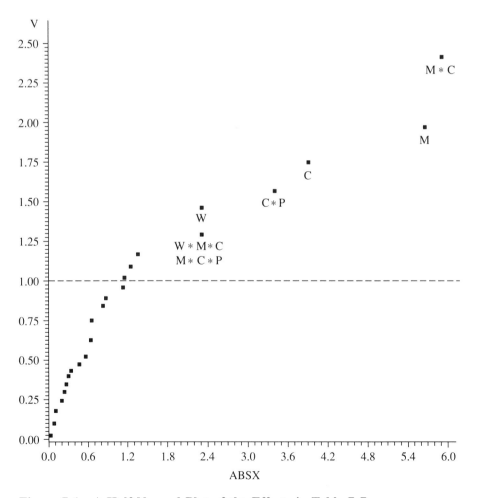

**Figure 7.1   A Half-Normal Plot of the Effects in Table 7.7.**

For example, the data in Table 7.5 were analyzed with SAS®
ANOVA using the statements:

```
PROC  ANOVA;
CLASSES  W  M  T  C  P;
MODEL  QUALITY  =  W  M  W*M  T  W*T  M*T
W*M*T  C  W*C  M*C  W*M*C  T*C  W*T*C  M*T*C
W*M*T*C  P  W*P  M*P  W*M*P  T*P  W*T*P  M*T*P
W*M*T*P  C*P  W*C*P  M*C*P  W*M*C*P  T*C*P
W*T*C*P  M*T*C*P  W*M*T*C*P;
```

The resulting analysis of variance table is shown in Table 7.7.

Notice that there are no $F$-values given or significance levels printed
in Table 7.7. This is because there is no estimate of $\sigma^2$ available from
this analysis. Notice also that the printed value of the ERROR MEAN
SQUARE is .00000000 and is based on 0 degrees of freedom.

For the next step in the analysis, the sums of squares in Table 7.7
corresponding to each of the effects in the model statement were placed
in a data file and a half-normal plot was obtained using SASGRAPH®
with the following statements:

```
GOPTIONS  HSIZE=7  VSIZE=7;

DATA  SUMSQRS;
INPUT  EFFECT  $  1-10  SS;
X  =  SQRT(SS);
ABSX  =  ABS(X);
N  =  31;
CARDS;
W                5.44500
M               31.60125
W*M              0.72000
  .                  .
  .                  .
  .                  .
W*M*T*C*P        0.10125

PROC  RANK  OUT=RANKS;
VAR  ABSX;
RANKS  R;

DATA  PLOTDATA;
SET  RANKS;
RSTAR=(R-.5)/N;
P=RSTAR/2+.5;
V=PROBIT(P);
PROC  GPLOT  DATA=PLOTDATA;
PLOT  V*ABSX/VZERO  HZERO  VREF=1  CVREF=RED;
```

The SASGRAPH® plot obtained from the above commands is shown in Figure 7.1.

Examination of the plot in Figure 7.1 reveals that the effects which appear to be significant are: $M*C$, $M$, $C$, $C*P$, $W$, $W*M*C$, and $M*C*P$. These are the same effects which were discovered to be significant when the data were analyzed by the analysis of variance procedure in Section 7.4.1.

To estimate $\sigma$, one needs to remove these seven effects from the data and then construct a new half-normal plot of the remaining effects. This can be done by once again using the above SASGRAPH® statements after replacing the line N = 31; with the line N = 24; and removing the seven largest effect sums of squares from the data. The resulting plot is given in Figure 7.2. The plotted points in Figure 7.2 appear to fall on a straight line, indicating that there are no more significant effects. The abscissa of the intersection of a line through the plotted points and the horizontal line through 1.0 is the estimate of $\sigma$. Thus the half-normal technique gives $\hat{\sigma} = .62$, which agrees extremely well with .61, the value which was obtained from pooling the high-order interaction sum of squares.

**7.5
COMPLETING
THE
ANALYSIS**

To complete the analysis of these data, first note that none of the effects involving factor $T$ (temperature) was significant. This indicates that temperatures in the range 300 to 500°F will have little, if any, effect on cake quality. Next note that factor $P$ (mixer type) is involved in two of the significant interactions, namely $P*C$ and $P*C*M$. This means that optimum levels of cooking oil and mixing time will be different for each mixer type. Thus, the easiest way to make sense out of these data might be to sort the data by the level of factor $P$ and to study the effects of $W$, $M$, and $C$ at each level of factor $P$. A sum of squares corresponding to each of these effects can be obtained by using the following SAS® commands:

```
PROC  SORT;  BY  P;

PROC  ANOVA;  BY  P;
CLASSES  W  M  P;
MODEL  QUALITY  =  W  M  W*M  C  W*C  M*C  W*M*C;
```

Factor $T$ was not included in the above analysis statements because it was not included in any of the significant effects in the first analysis. These analyses will only be used to compute sums of squares corresponding to each of the effects in the model. The significance of these effects will be determined by computing $F$-values by dividing each of the sums of squares by the estimate of $\sigma^2$ obtained from either of the analyses in Section 7.4.1 or 7.4.2. The results from these two analyses are shown in Tables 7.8 and 7.9.

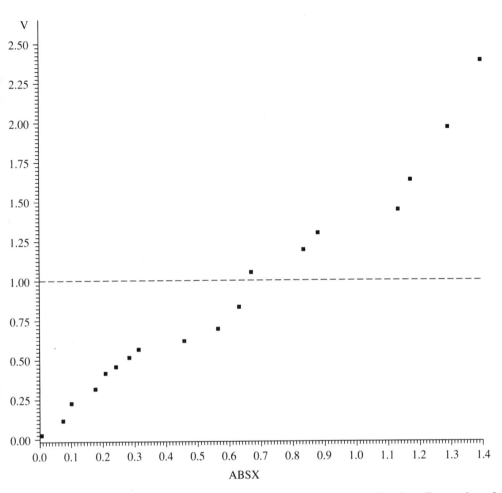

**Figure 7.2   A Half-Normal Plot of the Effects in Table 7.7 After Removing the Seven Largest Effects.**

New $F$-values are computed for each of the effects in Tables 7.8 and 7.9 by dividing each of the sums of squares by $\hat{\sigma}^2 = (.62)^2 = .3844$. The significant effects in Table 7.8 are $W$, $M$, $C$, $M*C$, and $W*M*C$ with new $F$-values equal to 17.93, 33.25, 68.32, 86.76, and 15.30, respectively. The significant effects in Table 7.9 are $M$ and $M*C$ with new $F$-values equal to 49.79 and 16.59, respectively. For mixer type $P_1$, the estimates of the expected responses to the treatments under study are shown in Table 7.10. The estimates in Table 7.10 are based on the means of two observations, and hence, an LSD for comparing pairs of treatment combination means in Table 7.10 is

$$\text{LSD}_{.05} = t_{.025} \cdot \hat{\sigma} \cdot \sqrt{2/r} = 2 \cdot .62 \cdot \sqrt{2/2} = 1.24$$

**Table 7.8   Analysis of the Data in Table 7.5 for Mixer Type $P_1$**

$P = 0$

*Analysis of Variance Procedure*

Dependent Variable: Quality

| Source | DF | Sum of squares | | Mean square | F value |
|---|---|---|---|---|---|
| Model | 7 | 86.29937500 | | 12.32848214 | 36.06 |
| Error | 8 | 2.73500000 | | 0.34187500 | PR > F |
| Corrected Total | 15 | 89.03437500 | | | 0.0001 |

| R-square | C.V. | Root MSE | | Quality mean | |
|---|---|---|---|---|---|
| 0.969282 | 11.8571 | 0.58470078 | | 4.93125000 | |

| Source | DF | ANOVA SS | F value | PR > F | |
|---|---|---|---|---|---|
| W | 1 | 6.89062500 | 20.16 | 0.0020 | |
| M | 1 | 12.78062500 | 37.38 | 0.0003 | |
| W*M | 1 | 0.85562500 | 2.50 | 0.1523 | |
| C | 1 | 26.26562500 | 76.83 | 0.0001 | |
| W*C | 1 | 0.27562500 | 0.81 | 0.3955 | |
| M*C | 1 | 33.35062500 | 97.55 | 0.0001 | |
| W*M*C | 1 | 5.88062500 | 17.20 | 0.0032 | |

**Table 7.9   Analysis of the Data in Table 7.5 for Mixer Type $P_2$**

$P = 1$

*Analysis of Variance Procedure*

Dependent Variable: Quality

| Source | DF | Sum of squares | | Mean square | F value |
|---|---|---|---|---|---|
| Model | 7 | 27.04937500 | | 3.86419643 | 10.36 |
| Error | 8 | 2.98500000 | | 0.37312500 | PR > F |
| Corrected Total | 15 | 30.03437500 | | | 0.0019 |

| R-square | C.V. | Root MSE | | Quality mean | |
|---|---|---|---|---|---|
| 0.900614 | 12.1409 | 0.61083959 | | 5.03125000 | |

| Source | DF | ANOVA SS | F value | PR > F | |
|---|---|---|---|---|---|
| W | 1 | 0.45562500 | 1.22 | 0.3013 | |
| M | 1 | 19.14062500 | 51.30 | 0.0001 | |
| W*M | 1 | 0.07562500 | 0.20 | 0.6645 | |
| C | 1 | 0.14062500 | 0.38 | 0.5563 | |
| W*C | 1 | 0.18062500 | 0.48 | 0.5063 | |
| M*C | 1 | 6.37562500 | 17.09 | 0.0033 | |
| W*M*C | 1 | 0.68062500 | 1.82 | 0.2138 | |

**Table 7.10   Estimates for Mixer Type = $P_1$**

| W | M | C | Estimate |
|---|---|---|----------|
| 0 | 0 | 0 | 4.35 |
| 0 | 0 | 1 | 5.50 |
| 0 | 1 | 0 | 4.00 |
| 0 | 1 | 1 | 8.50 |
| 1 | 0 | 0 | 4.05 |
| 1 | 0 | 1 | 2.25 |
| 1 | 1 | 0 | 2.20 |
| 1 | 1 | 1 | 8.60 |

Examination of the results in Table 7.10 shows that cakes of the highest quality can be made by using the higher level of mixing time and the higher amount of cooking oil, and for this combination of mixing time and cooking oil, the amount of water used has no real affect. The amount of water used does have a significant affect on cake quality when mixing time is at the low level and cooking oil is at a high level. The amount of water used also has a significant effect on cake quality when mixing time is at a high level and cooking oil is at a low level. In both instances the higher amount of water gives a much lower-quality cake. When both mixing time and cooking oil are at their lowest levels, the amount of water used has little affect on cake quality.

For mixer type $P_2$, one should examine the $M*C$ treatment combination means as this was the only significant interaction for this mixer type. Table 7.11 shows the estimates of the $M*C$ treatment combination means for mixer type $P_2$. The estimates in Table 7.11 are based on the means of four observations, and hence an LSD for comparing pairs of treatment combination means in Table 7.11 is

$$\text{LSD}_{.05} = t_{.025}\hat{\sigma} \cdot \sqrt{2/r} = 2 \cdot .62 \cdot \sqrt{2/4} = .88$$

**Table 7.11   Estimates for Mixer Type = $P_2$**

| M | C | Estimate |
|---|---|----------|
| 0 | 0 | 4.475 |
| 0 | 1 | 3.40 |
| 1 | 0 | 5.40 |
| 1 | 1 | 6.85 |

Examination of the results in Table 7.11 show that cakes of the highest quality can be made by using the higher level of mixing time and the higher amount of cooking oil, and for this combination of mixing time and cooking oil, the amount of water used once again has no real affect since the main effect for water was not significant in the analysis of variance table. For mixer type $P_2$, the amount of water used does not have a significant affect on cake quality.

In conclusion, if one were going to conduct an experiment to further refine the process, it will be necessary to conduct separate experiments for each mixer type. In both of these experiments, water level and temperature could be chosen at some intermediate level as long as cooking oil and mixing times are chosen to have relatively high levels. The techniques which will be discussed in Chapter 11 may be quite helpful in fine-tuning the process.

# 8

# Blocking $2^n$ Factorial Treatment Structures

In Chapter 7 it was noted that $2^n$ experiments which involve large numbers of factors have so many high-order interaction effects that they generally provide more information for estimating the experimental error variance than is probably necessary. This makes it possible to conduct these kinds of factorial experiments more efficiently, by grouping the treatment combinations into subgroups in special ways, and then randomly assigning the treatments in each subgroup to a blocked set of experimental units. Methods required for utilizing blocking in $2^n$ factorial experiments is the subject of this chapter.

## 8.1 A $2^3$ EXPERIMENT IN TWO BLOCKS OF SIZE 4

Blocking in $2^n$ factorial experiments will first be illustrated when $n = 3$, although blocking cannot generally be recommended for factorial experiments of this size unless one conducts more than one replication of each treatment combination. From the last line of Table 7.2, note that the contrast of the treatment combinations which measures the three-factor interaction is

$$A*B*C = [-(1) + a + b - ab + c - ac - bc + abc]$$

Suppose one assigns the eight treatment combinations to two blocks of four experimental units each so that all of those treatment combinations in the $A*B*C$ interaction contrast which are preceded by $+$ signs go into one block and all of those treatment combinations which are preceded by $-$ signs go into the other block. Table 8.1 shows the experimental plan for conducting a $2^3$ experiment as described above.

It is interesting to note that block 1 contains all of those treatment combinations for which the sum of the three factor levels is odd and block 2 contains all of those treatment combinations for which the sum of the three factor levels is even.

What are the advantages and disadvantages of running a $2^3$ factorial experiment in two blocks of size 4? The only disadvantage for the experimental plan described in Table 8.1 is that one is no longer able to

### Table 8.1   A $2^3$ Experiment in Two Blocks of Size 4

| BLOCK 1 TREATMENT COMBINATION | | | | | BLOCK 2 TREATMENT COMBINATION | | | | |
|---|---|---|---|---|---|---|---|---|---|
| $A$ | $B$ | $C$ | | | $A$ | $B$ | $C$ | | |
| 1 | 0 | 0 | : | $a$ | 0 | 0 | 0 | : | (1) |
| 0 | 1 | 0 | : | $b$ | 1 | 1 | 0 | : | $ab$ |
| 0 | 0 | 1 | : | $c$ | 1 | 0 | 1 | : | $ac$ |
| 1 | 1 | 1 | : | $abc$ | 0 | 1 | 1 | : | $bc$ |

accurately measure the three-factor interaction. This is because the contrast of the eight treatment combinations which measures the $A * B * C$ interaction is the same contrast as the one which would measure the Block effect (i.e., the sum of all treatment combinations in block 1 minus the sum of all treatment combinations in block 2). It is said that the $A * B * C$ interaction is *confounded* with Blocks. That is, one is not able to distinguish between block effects and the $A * B * C$ interaction effect. Of course, the hope is that there is no three-factor interaction, in which case it would not be necessary to estimate it. Such hope may be suspect for three-factor interactions, but experience has shown that four-factor and higher-order interactions are rarely present and one loses very little information by confounding such high-order interaction effects with blocks.

The primary advantage of conducting a $2^3$ experiment according to the plan in Table 8.1 when there is no real three-factor interaction is that one is still able to measure all of the main effects and all two-factor interactions. Furthermore, one is usually able to measure these effects more accurately since the experimental error variance is usually reduced by blocking the experimental units into subgroups of more homogeneous units.

To see that all main effects and two-factor interactions can be estimated freely of the block effects (i.e., these effects are not confounded with blocks), consider, for example, the contrast which measures the $A$ effect. This contrast is

$$A = [-(1) + a - b + ab - c + ac - bc + abc]$$

In this contrast two of the treatment combinations in block 1 have + signs ($a$ and $abc$) and the other two treatment combinations in block 1 have − signs ($b$ and $c$). Thus the effect of block 1 cancels out of the $A$ contrast. Similarly, in the $A$ contrast, two of the treatment combinations in block 2 have + signs ($ab$ and $ac$) while the other two have − signs [(1) and $bc$], and as a result the effect of block 2 also cancels out of the $A$ contrast. This same pattern holds true for each of the other main effect contrasts and for each of the two-factor interaction contrasts, and hence, each of these effects is free of (not confounded with) block effects.

Blocking $2^n$ factorial experiments into two blocks of size $2^{n-1}$ is a straightforward generalization of the results presented in the last section. The effect which should be confounded with blocks is the highest-order interaction. This is the effect which is most likely to be nonexistent. To block in this manner, all of those treatment combinations which have the sum of their factor levels odd should go into one block and those which have the sum of their factor levels even should go into the other block.

For example, consider the $2^5$ experiment described in Section 7.4. Suppose two scientists are going to share the work required in conducting

**8.2
$2^n$
FACTORIAL
EXPERIMENTS
IN TWO
BLOCKS
OF SIZE $2^{n-1}$**

this experiment, with each scientist performing one-half of the required 32 treatment combinations. To reduce the variability in the experiment which might be caused by the two scientists, it is decided to block on scientists. Thus one scientist will perform those experiments which have the sum of their factor levels odd and the other scientist will perform those experiments which have the sum of their factor levels even. Those treatment combinations which should be performed by each scientist are illustrated in Table 8.2.

**Table 8.2   A $2^5$ Experiment in Two Blocks of Size 16**

|           | TREATMENT COMBINATION | | | | |
|-----------|---|---|---|---|---|
| *Scientist* | *W* | *M* | *T* | *C* | *P* |
| 1 | 1 | 0 | 0 | 0 | 0 |
| 1 | 0 | 1 | 0 | 0 | 0 |
| 1 | 0 | 0 | 1 | 0 | 0 |
| 1 | 0 | 0 | 0 | 1 | 0 |
| 1 | 0 | 0 | 0 | 0 | 1 |
| 1 | 1 | 1 | 1 | 0 | 0 |
| 1 | 1 | 1 | 0 | 1 | 0 |
| 1 | 1 | 1 | 0 | 0 | 1 |
| 1 | 1 | 0 | 1 | 1 | 0 |
| 1 | 1 | 0 | 1 | 0 | 1 |
| 1 | 1 | 0 | 0 | 1 | 1 |
| 1 | 0 | 1 | 1 | 1 | 0 |
| 1 | 0 | 1 | 1 | 0 | 1 |
| 1 | 0 | 1 | 0 | 1 | 1 |
| 1 | 0 | 0 | 1 | 1 | 1 |
| 1 | 1 | 1 | 1 | 1 | 1 |
| 2 | 0 | 0 | 0 | 0 | 0 |
| 2 | 1 | 1 | 0 | 0 | 0 |
| 2 | 1 | 0 | 1 | 0 | 0 |
| 2 | 1 | 0 | 0 | 1 | 0 |
| 2 | 1 | 0 | 0 | 0 | 1 |
| 2 | 0 | 1 | 1 | 0 | 0 |
| 2 | 0 | 1 | 0 | 1 | 0 |
| 2 | 0 | 1 | 0 | 0 | 1 |
| 2 | 0 | 0 | 1 | 1 | 0 |
| 2 | 0 | 0 | 1 | 0 | 1 |
| 2 | 0 | 0 | 0 | 1 | 1 |
| 2 | 1 | 1 | 1 | 1 | 0 |
| 2 | 1 | 1 | 1 | 0 | 1 |
| 2 | 1 | 1 | 0 | 1 | 1 |
| 2 | 1 | 0 | 1 | 1 | 1 |
| 2 | 0 | 1 | 1 | 1 | 1 |

Each scientist should randomize the order in which their selected set of treatment combinations is performed in order to eliminate or reduce any possible biases in the experimental results that might be associated with the time at which the experiment is performed.

Unfortunately, some scientists might divide the treatment combinations in some other manner. For example, it would be much more convenient to let one scientist conduct that half of the treatment combinations which requires mixer type $P_1$ and the other scientist conduct that half which requires mixer type $P_2$. That is, one scientist does those runs in Table 7.5 which have factor $P$ at an odd level and the other scientist does those runs which have factor $P$ at an even level. If this experiment were conducted in this way, the main effect of factor $P$ would be confounded with the Scientist effect, and if this effect were observed to be large, which is quite likely, one would not be able to tell if the significance of this effect is due to the different scientists or to the different mixer types. If one is going to confound Scientist effects with Treatment effects, one should always try to confound the Scientist effect with that Treatment effect which is most likely to be nonexistent, in this case, the five-factor interaction.

Other researchers might divide the two sets of treatment combinations in some random manner. This would be extremely unfortunate, for if this were done and if there were a significant effect due to the different scientists, then almost all of the effects of interest (main effects, two-factor interactions, etc.) are likely to be confounded with Scientists. In this case, no reliable information would be available from the data collected, and those resources used to collect the data would be wasted.

## 8.3 ANALYZING $2^n$ FACTORIAL EXPERIMENTS CONDUCTED IN TWO BLOCKS OF SIZE $2^{n-1}$

Analyzing $2^n$ factorial experiments which have been conducted in two blocks of size $2^{n-1}$ is no more difficult than analyzing $2^n$ experiments in which blocking has not been used. Both the analysis of variance method and half-normal plot method discussed in Chapter 7 can be used with some slight alterations. The alterations required for the analysis of variance method will be discussed in Section 8.3.1, while those required for the half-normal plot method will be discussed in Section 8.3.2.

### 8.3.1 The Analysis of Variance Method with Blocking

This method can be used whenever the experiment is large enough so that the high-order interaction effects are sufficient in number to be able to adequately measure the experimental error variance by pooling them together. This is the case whenever $n > 5$. The data can be analyzed by using any statistical computing package which will analyze $n$-way cross-classified experiments. The model statement which should be used will include a term for Blocks plus terms for all main effects, two-factor interactions, and three-factor interactions. Those effects not included in the model, namely, the four-factor, five-factor, . . . , $(n - 1)$-factor interactions will automatically be pooled to provide an estimate of the experimental error variance. The $n$-factor interaction term should not be included in the model statement. If it were included, the sum of squares corresponding to this effect would be equal to the Block sum of squares which has already been removed from the Error sum of squares. Putting both terms in the model would remove this effect twice (at least for many statistical computing routines), and hence could lead to erroneous results. In fact, some packages may produce an Error sum of squares which is negative when both terms are included in the model.

To further illustrate, if the experiment described in Section 7.4 were conducted according to the plan in Table 8.2, the SAS® ANOVA statements required for an analysis using the analysis of variance method would be

```
PROC ANOVA;
CLASSES BLOCK W M T C P;
MODEL QUALITY = BLOCK W M W*M T W*T M*T
                W*M*T C W*C M*C W*M*C T*C
                W*T*C M*T*C P W*P M*P W*M*P
                T*P W*T*P M*T*P C*P W*C*P
                M*C*P T*C*P;
```

For this experiment there would be 5 degrees of freedom for estimating the experimental error variance, and the estimate of the experimental error variance is obtained by pooling the sums of squares corresponding to each of the five four-factor interactions.

**8.3.2**
**The Half-Normal Plot Method with Blocking**

If one is going to use the half-normal plot method to analyze $2^n$ factorial experiments which have been conducted in two blocks of size $2^{n-1}$, there are no adjustments which need to be made. That is, one would still fit a model which includes all of the main effects, two-factor interactions, . . . , $n$-factor interaction, so that a single-degree-of-freedom sum of squares is computed for each effect. One just needs to remember that the $n$-factor interaction is confounded with blocks, and if this effect is significantly large, it most likely means that there is a significant difference between the two blocks of experimental units and not that there is a significant $n$-factor interaction.

**8.4**
**CONFOUNDING $2^n$ FACTORIAL EXPERIMENTS IN MORE THAN TWO BLOCKS**

In the preceding sections of this chapter, the design and analysis of $2^n$ factorial experiments conducted in two blocks of size $2^{n-1}$ was considered. It is possible to subdivide the $2^n$ treatment combinations into more blocks of smaller sizes than was required in the previous sections. The advantage of having smaller block sizes is that the experimental units which are grouped into smaller blocks can be more homogeneous, and as a result, such experiments will usually provide smaller estimates of the experimental error variance. In this case more accurate comparisons of important effects will be possible. The disadvantage is that in order to conduct such experiments, additional effects must also be confounded with blocks. This is not a major disadvantage, if only higher-order interaction effects are confounded with blocks, or if effects which are known, by prior experience, not to be real are confounded with blocks.

If one is going to wisely conduct a $2^n$ experiment in $2^k$ blocks each containing $2^{n-k}$ experimental units, then there will be $2^k - 1$ effects which will be confounded with blocks. If one conducts such an experiment unwisely, all effects may be confounded with blocks. The minimum number of effects which must be confounded with blocks is $2^k - 1$. For example, if one is going to conduct a $2^5$ experiment in four blocks with each block containing eight experimental units, there will be a minimum of three effects which must be confounded with blocks. One is able to specify two of these three effects, but the third effect will be determined by the choice of these first two. In general, to conduct a $2^n$ experiment in $2^k$ blocks, one must be able to specify $k$ effects to confound with blocks, and these $k$ effects will determine an additional $2^k - k - 1$ effects, which will also be confounded with blocks.

To illustrate, suppose one desires to conduct a $2^4$ factorial experiment in $2^2 = 4$ blocks of size $2^{4-2} = 4$ experimental units each. How should one divide the 16 treatment combinations into groups if one wishes to confound the four-factor interaction effect and one of the two-factor interaction effects with blocks? And in addition, what is the third effect which will also be confounded with blocks?

The notation that will be used in this section is the same notation that was used in Table 7.3. Suppose in this $2^4$ experiment one wishes to confound the $A * B * C * D$ interaction and the $A * B$ interaction with blocks. From Table 7.3 one can see that the $A * B * C * D$ interaction is defined by

$$A * B * C * D = (1) - a - b + ab - c + ac + bc - abc$$
$$-d + ad + bd - abd + cd - acd - bcd + abcd$$

and the $A * B$ interaction is defined by

$$A * B = (1) - a - b + ab + c - ac - bc + abc + d - ad - bd + abd$$
$$+ cd - acd - bcd + abcd$$

Both of these effects will be confounded with blocks if one places all of those treatment combinations which have $+$ signs in both contrasts in block 1, those which have $+$ signs in the first contrast and $-$ signs in the second contrast in block 2, those which have $-$ signs in the first contrast and $+$ signs in the second contrast in block 3, and those contrasts which have $-$ signs in both contrasts in block 4. Upon doing this, the treatments which fall into each block are shown in Table 8.3.

### Table 8.3   A $2^4$ Experiment in Four Blocks of Size 4

| | BLOCK 1 TRT. COMB. | | | | | BLOCK 2 TRT. COMB. | | | | | BLOCK 3 TRT. COMB. | | | | | BLOCK 4 TRT. COMB. | | | |
|---|---|---|---|---|---|---|---|---|---|---|---|---|---|---|---|---|---|---|---|
| | $A$ | $B$ | $C$ | $D$ | | $A$ | $B$ | $C$ | $D$ | | $A$ | $B$ | $C$ | $D$ | | $A$ | $B$ | $C$ | $D$ |
| (1) | 0 | 0 | 0 | 0 | $ac$ | 1 | 0 | 1 | 0 | $c$ | 0 | 0 | 1 | 0 | $a$ | 1 | 0 | 0 | 0 |
| $ab$ | 1 | 1 | 0 | 0 | $ad$ | 1 | 0 | 0 | 1 | $d$ | 0 | 0 | 0 | 1 | $b$ | 0 | 1 | 0 | 0 |
| $cd$ | 0 | 0 | 1 | 1 | $bc$ | 0 | 1 | 1 | 0 | $abc$ | 1 | 1 | 1 | 0 | $acd$ | 1 | 0 | 1 | 1 |
| $abcd$ | 1 | 1 | 1 | 1 | $bd$ | 0 | 1 | 0 | 1 | $abd$ | 1 | 1 | 0 | 1 | $bcd$ | 0 | 1 | 1 | 1 |
| | Both + | | | | | + and − | | | | | − and + | | | | | Both − | | | |

Next let $\ell(F)$ represent the level of a factor $F$. Also let

$$L_1 = \ell(A) + \ell(B) + \ell(C) + \ell(D)$$
$$L_2 = \ell(A) + \ell(B)$$

$L_1$ corresponds to the $A*B*C*D$ interaction and $L_2$ corresponds to the $A*B$ interaction. Note that block 1 contains those treatment combinations for which both $L_1$ and $L_2$ are even, block 2 contains those treatment combinations for which $L_1$ is even and $L_2$ is odd, block 3 contains those treatment combinations for which $L_1$ is odd and $L_2$ is even, and block 4 contains those treatment combinations for which both $L_1$ and $L_2$ are odd. Note also that specifying $L_1$ to be odd or even and specifying $L_2$ to be odd or even forces the oddness or evenness of $L_3 = \ell(C) + \ell(D)$. That is, if $L_1$ is even and $L_2$ is even, then $L_3$ must also be even. Similarly, if $L_1$ is even and $L_2$ is odd, then $L_3$ must be odd, and if $L_1$ is odd and $L_2$ is even, then $L_3$ must be odd, and finally, if both $L_1$ and $L_2$ are odd, then $L_3$ must be even. This would seem to imply that if one confounds both the $A*B*C*D$ interaction and the $A*B$ interaction with blocks, then the $C*D$ interaction will also be confounded with blocks. This is in fact true. To see this note that the $C*D$ interaction is defined by

$$C*D = (1) + a + b + ab - c - ac - bc - abc - d - ad$$
$$-bd - abd + cd + acd + bcd + abcd$$

and all of the treatment combinations which fall in blocks 1 and 4 have + signs for this contrast, while all of those in blocks 2 and 3 have − signs for this contrast. Hence the $C*D$ interaction contrast also contrasts blocks 1 and 4 versus blocks 2 and 3. Thus the $C*D$ interaction contrast also measures Block effects, and therefore the $C*D$ interaction effect is also confounded with blocks.

If one knows the effects used to assign treatment combinations to blocks, one can more easily determine the additional effects which are also confounded with blocks by finding what are called the *generalized interactions* between all subsets of the defining effects. The generalized interaction of two or more effects is found by combining all of the letters that appear in these effects and cancelling out those that occur an even number of times. For example, the generalized interaction between $A * B * C * D$ and $A * B$ is

$$(A * B * C * D) * (A * B) = C * D$$

as $A$ occurs twice and $B$ occurs twice and hence they are cancelled out.

Suppose that one decides to assign the treatment combinations in a $2^4$ experiment to four blocks by confounding the $A * B * C * D$ interaction and the $A * B * C$ interaction. In this case the other effect which would also be confounded with blocks is

$$(A * B * C * D) * (A * B * C) = D$$

Hence confounding the four-factor interaction and a three-factor interaction in a $2^4$ experiment also confounds a main effect with blocks. In nonreplicated experiments, one would never want to confound a main effect with blocks, so that an assignment of treatment combinations to blocks such as described above would never be desirable.

As a final example, consider designing a $2^6$ factorial experiment into $2^3 = 8$ blocks of $2^3 = 8$ experimental units each. To assign the treatment combinations to blocks, suppose one decides to confound the $A*B*C*D$, $A * B * E * F$, and $A * C * E$ interaction effects with blocks. Let

$$L_1 = \ell(A) + \ell(B) + \ell(C) + \ell(D)$$
$$L_2 = \ell(A) + \ell(B) + \ell(E) + \ell(F)$$
$$L_3 = \ell(A) + \ell(C) + \ell(E)$$

Table 8.4 describes which treatment combinations should be assigned to each block. For example, block 1 would contain all of those

**Table 8.4   Rules for Assigning the Treatment Combinations in a $2^6$ Factorial Experiment to Eight Blocks of Size 8**

|        | Block 1 | Block 2 | Block 3 | Block 4 | Block 5 | Block 6 | Block 7 | Block 8 |
|--------|---------|---------|---------|---------|---------|---------|---------|---------|
| $L_1$: | Even    | Even    | Even    | Even    | Odd     | Odd     | Odd     | Odd     |
| $L_2$: | Even    | Even    | Odd     | Odd     | Even    | Even    | Odd     | Odd     |
| $L_3$: | Even    | Odd     | Even    | Odd     | Even    | Odd     | Even    | Odd     |

treatment combinations for which $L_1$, $L_2$, and $L_3$ are all even, and block 2 would contain all of those treatment combinations for which $L_1$ and $L_2$ are even and $L_3$ is odd, etc. Table 8.5 shows the actual assignment of the treatment combinations to the blocks. Of course, when conducting the experiment, the blocks in Table 8.5 should be randomly assigned to the blocks of experimental units and the treatment combinations within a block should be randomly assigned to the experimental units within the block of experimental units.

To determine the other effects which would also be confounded with blocks for the assignment of treatment combinations to blocks given in Table 8.5, one must simplify the generalized interactions of all subsets of the defining effects. The generalized interactions of all subsets of the defining effects are

$$(A * B * C * D) * (A * B * E * F) = C * D * E * F$$

$$(A * B * C * D) * (A * C * E) = B * D * E$$

$$(A * B * E * F) * (A * C * E) = B * C * F$$

$$(A * B * C * D) * (A * B * E * F) * (A * C * E) = A * D * F$$

Hence those effects which are confounded with blocks for the experimental plan in Table 8.5 are the following seven effects: $A * B * C * D$, $A * B * E * F$, $A * C * E$, $C * D * E * F$, $B * D * E$, $B * C * F$, and $A * D * F$.

Cochran and Cox (1957) give several experimental plans for confounding effects in $2^n$ factorial experiments with blocks at the end of their Chapter 6. Each replicate in their plans confounds a different set of effects with blocks. These plans can also be used for replicated experiments. To see how this might be done see Cochran and Cox (1957); Box, Hunter, and Hunter (1978); or Hicks (1982).

**Table 8.5   Treatment Combination Assignment for a $2^6$ Factorial Experiment in Eight Blocks of Size 8**

| Block 1 | Block 2 | Block 3 | Block 4 | Block 5 | Block 6 | Block 7 | Block 8 |
|---------|---------|---------|---------|---------|---------|---------|---------|
| (1)     | ab      | ac      | bc      | abc     | c       | b       | a       |
| abcd    | cd      | bd      | ad      | d       | abd     | acd     | bcd     |
| bce     | ace     | abe     | e       | ae      | be      | ce      | abce    |
| ade     | bde     | cde     | abcde   | bcde    | acde    | abde    | de      |
| acf     | bcf     | f       | abf     | bf      | af      | abcf    | cf      |
| bdf     | adf     | abcdf   | cdf     | acdf    | bcdf    | df      | abdf    |
| abef    | ef      | bcef    | acef    | cef     | abcef   | aef     | bef     |
| cdef    | abcdef  | adef    | bdef    | abdef   | def     | bcdef   | acdef   |

The analysis of $2^n$ factorial experiments in which the treatment combinations have been assigned to more than two blocks is considered in the next section.

**8.5**
**ANALYZING CON-FOUNDED $2^n$ FACTORIAL EXPERIMENTS WHEN EXPERIMEN-TAL UNITS HAVE BEEN GROUPED INTO MORE THAN TWO BLOCKS**

In this section, the analysis of the $2^6$ experiment with the blocking structure given in Table 8.5 will be discussed. Such experiments can be analyzed by both the analysis of variance method and the half-normal plot method. The analysis of variance method will be discussed here and the half-normal plot method will be discussed in Section 8.5.1.

To analyze data arising from the experimental plan in Table 8.5 using the analysis of variance method, one would fit a model which has a term in it for Blocks as well as terms in it for all main effects, all two-factor interactions, and all three-factor interactions except those which are confounded with blocks. It is important to be sure not to include any terms in the model that are confounded with blocks. Doing so would produce an inaccurate estimate of the experimental error variance with most statistical computing packages and, as a result, erroneous statistical tests. Thus it is extremely important to be able to determine which effects are confounded with blocks in these types of experimental plans.

If one were to analyze data arising from the experimental plan in Table 8.5 using SAS®, the statements which would be required are

```
PROC ANOVA;
CLASSES BLOCKS  A B C D E F;
MODEL dep. vars. = BLOCKS A B A *B C
              A*C B*C A*B*C D
              A*D B*D A*B*D C*D
              A*C*D B*C*D E A*E B*E
              A*B*E C*E B*C*E D*E
              A*D*E C*D*E F A*F
              B*F A*B*F C*F A*C*F
              D*F B*D*F C*D*F E*F
              A*E*F B*E*F
              C*E*F D*E*F;
```

Note that the above model statement does not include terms for the $A * C * E$, $B * D * E$, $B * C * F$, and $A * D * F$ interactions, which are the three-factor interactions which were confounded with blocks in the experimental plan. This experimental plan gives 19 degrees of freedom for estimation of the experimental error variance.

In quite a few experimental situations, three-factor interactions also tend to be nonexistent or, at least, very small when compared to main effects and two-factor interactions. If this is the case, then all three-factor interactions which are not confounded with blocks could also be pooled into the estimate of the experimental error variance. For the experimental plan in Table 8.5, pooling the remaining three-factor interactions into the estimate of experimental error would give an additional 16 degrees of freedom for estimating the experimental error variance, for a total of 35 degrees of freedom.

**8.5.1**
**The Half-Normal Plot Method in Confounded $2^n$ Experiments**

If one is going to use the half-normal plot method to analyze $2^n$ factorial experiments which have been conducted in $2^k$ blocks of size $2^{n-k}$, there are, once again, no adjustments which need to be made. That is, one would still fit a model which includes all of the main effects, two-factor interactions, . . . , $n$-factor interaction, so that a single-degree-of-freedom sum of squares is computed for each effect. One just needs to remember which of the effects are confounded with blocks, and if such an effect is significantly large, it most likely means that there is a significant difference between the blocks of experimental units and not that this effect is significant.

Both the analysis of variance method and the half-normal plot method will be illustrated by an example in the next section.

**8.6**
**AN EXAMPLE**

A $2^6$ factorial experiment was conducted to improve the quality of a frozen food product. The factors which were thought to influence product quality were mix temperature, mixing speed, freezing temperature, air velocity, end product temperature, and package type. The factors to be studied and their levels are described in Table 8.6. A complete replication of this experiment requires 64 runs. Since the raw materials used in making the product could be changing from day to day, it was decided that it might be wise to block on days in order to reduce experimental variability when conducting the experiment; this should enable one to get more accurate information from the experiment. The maximum number of runs which could be performed on any given day was equal to 8. Thus

**Table 8.6   Definition of the Treatment Structure in a $2^6$ Factorial Experiment to Improve the Quality of a Frozen Food Product**

| | LEVEL | |
|---|---|---|
| *Variable* | *Low* | *High* |
| MT: mixing time | 3 min | 6 min |
| MS: mixing speed | 75 rpm | 150 rpm |
| FT: freezing temperature | $-40°F$ | $-20°F$ |
| AV: air velocity | 100 fpm | 300 fpm |
| ET: end product temperature | $0°F$ | $10°F$ |
| PT: package type | Rectangular | Round |

it was decided to conduct the $2^6$ factorial experiment in eight blocks of size 8.

The experimental plan described in Table 8.5 will be used to conduct this experiment. To ensure that one will be able to make valid inferences from the data collected, one should (1) randomly assign the factors in Table 8.6 to the symbols $A, B, C, D, E, F$ in Table 8.5, (2) randomly assign the blocks in Table 8.5 to the days on which the experiment is to be conducted, and (3) randomize the order of running the combinations within each block using a new randomization scheme within each block. The results of the first two of these randomizations are shown in Table 8.7; the run orders for each day are shown in Table 8.8, which also shows the results of the experiment.

The data in Table 8.8 were analyzed with SAS® ANOVA using the following statements:

```
PROC ANOVA;
CLASSES DAY MT MS FT AV ET PT;
MODEL QUALITY = DAY MT MS FT
                AV ET PT MT*MS MT*FT MT*AV MT*ET
                MT*PT MS*FT MS*AV MS*ET MS*PT FT*AV
                FT*ET FT*PT AV*ET AV*PT ET*PT
                MT*MS*FT MT*MS*AV MT*MS*ET
                MT*FT*ET MT*FT*PT MT*AV*ET
                MT*AV*PT MT*ET*PT MS*FT*AV
                MS*FT*ET MS*FT*PT
                MS*AV*PT MS*ET*PT FT*AV*ET
                FT*AV*PT AV*ET*PT;
```

The model statement used above includes terms for the blocking factor, Day, as well as terms for all main effects, all two-factor interactions, and those three-factor interactions which were not confounded with the blocking factor.

**Table 8.7  Random Assignment
of Factors to Symbols and Days to Blocks**

| Factor | Symbol | Day | Block |
|--------|--------|-----|-------|
| MT | B | 1 | 3 |
| MS | D | 2 | 8 |
| FT | C | 3 | 6 |
| AV | F | 4 | 1 |
| ET | A | 5 | 7 |
| PT | E | 6 | 2 |
|    |   | 7 | 5 |
|    |   | 8 | 4 |

**Table 8.8   Experimental Plan and Data for Improving the Quality
of a Frozen Food Product**

| Day | Run | ET | MT | FT | MS | PT | AV | Quality | Day | Run | ET | MT | FT | MS | PT | AV | Quality |
|---|---|---|---|---|---|---|---|---|---|---|---|---|---|---|---|---|---|
| 1 | 1 | 1 | 0 | 1 | 0 | 0 | 0 | 4.5 | 1 | 4 | 0 | 1 | 0 | 1 | 0 | 0 | 9.2 |
| 1 | 7 | 1 | 1 | 0 | 0 | 1 | 0 | 6.9 | 1 | 3 | 0 | 0 | 1 | 1 | 1 | 0 | 5.6 |
| 1 | 8 | 0 | 0 | 0 | 0 | 0 | 1 | 6.2 | 1 | 2 | 1 | 1 | 1 | 1 | 0 | 1 | 5.8 |
| 1 | 6 | 0 | 1 | 1 | 0 | 1 | 1 | 6.3 | 1 | 5 | 1 | 0 | 0 | 1 | 1 | 1 | 7.2 |
| 2 | 6 | 1 | 0 | 0 | 0 | 0 | 0 | 6.3 | 2 | 3 | 0 | 1 | 1 | 1 | 0 | 0 | 6.0 |
| 2 | 2 | 1 | 1 | 1 | 0 | 1 | 0 | 5.6 | 2 | 1 | 0 | 0 | 0 | 1 | 1 | 0 | 7.9 |
| 2 | 5 | 0 | 0 | 1 | 0 | 0 | 1 | 5.5 | 2 | 7 | 1 | 1 | 0 | 1 | 0 | 1 | 7.5 |
| 2 | 4 | 0 | 1 | 0 | 0 | 1 | 1 | 8.2 | 2 | 8 | 1 | 0 | 1 | 1 | 1 | 1 | 7.1 |
| 3 | 7 | 0 | 0 | 1 | 0 | 0 | 0 | 2.4 | 3 | 4 | 1 | 1 | 0 | 1 | 0 | 0 | 6.6 |
| 3 | 1 | 0 | 1 | 0 | 0 | 1 | 0 | 6.5 | 3 | 2 | 1 | 0 | 1 | 1 | 1 | 0 | 3.0 |
| 3 | 3 | 1 | 0 | 0 | 0 | 0 | 1 | 3.4 | 3 | 5 | 0 | 1 | 1 | 1 | 0 | 1 | 4.2 |
| 3 | 8 | 1 | 1 | 1 | 0 | 1 | 1 | 4.3 | 3 | 6 | 0 | 0 | 0 | 1 | 1 | 1 | 6.8 |
| 4 | 4 | 0 | 0 | 0 | 0 | 0 | 0 | 4.8 | 4 | 3 | 1 | 1 | 1 | 1 | 0 | 0 | 4.6 |
| 4 | 5 | 0 | 1 | 1 | 0 | 1 | 0 | 4.6 | 4 | 1 | 1 | 0 | 0 | 1 | 1 | 0 | 6.3 |
| 4 | 6 | 1 | 0 | 1 | 0 | 0 | 1 | 3.8 | 4 | 7 | 0 | 1 | 0 | 1 | 0 | 1 | 7.4 |
| 4 | 2 | 1 | 1 | 0 | 0 | 1 | 1 | 6.4 | 4 | 8 | 0 | 0 | 1 | 1 | 1 | 1 | 4.5 |
| 5 | 8 | 0 | 1 | 0 | 0 | 0 | 0 | 8.5 | 5 | 5 | 1 | 0 | 1 | 1 | 0 | 0 | 5.0 |
| 5 | 4 | 0 | 0 | 1 | 0 | 1 | 0 | 2.9 | 5 | 3 | 1 | 1 | 0 | 1 | 1 | 0 | 7.8 |
| 5 | 2 | 1 | 1 | 1 | 0 | 0 | 1 | 5.8 | 5 | 6 | 0 | 0 | 0 | 1 | 0 | 1 | 6.9 |
| 5 | 1 | 1 | 0 | 0 | 0 | 1 | 1 | 5.5 | 5 | 7 | 0 | 1 | 1 | 1 | 1 | 1 | 6.4 |
| 6 | 6 | 1 | 1 | 0 | 0 | 0 | 0 | 5.7 | 6 | 5 | 0 | 0 | 1 | 1 | 0 | 0 | 3.5 |
| 6 | 2 | 1 | 0 | 1 | 0 | 1 | 0 | 3.0 | 6 | 7 | 0 | 1 | 0 | 1 | 1 | 0 | 7.0 |
| 6 | 1 | 0 | 1 | 1 | 0 | 0 | 1 | 5.9 | 6 | 4 | 1 | 0 | 0 | 1 | 0 | 1 | 5.6 |
| 6 | 8 | 0 | 0 | 0 | 0 | 1 | 1 | 5.9 | 6 | 3 | 1 | 1 | 1 | 1 | 1 | 1 | 7.2 |
| 7 | 5 | 1 | 1 | 1 | 0 | 0 | 0 | 2.4 | 7 | 4 | 0 | 0 | 0 | 1 | 0 | 0 | 6.6 |
| 7 | 1 | 1 | 0 | 0 | 0 | 1 | 0 | 5.3 | 7 | 6 | 0 | 1 | 1 | 1 | 1 | 0 | 4.1 |
| 7 | 8 | 0 | 1 | 0 | 0 | 0 | 1 | 6.0 | 7 | 2 | 1 | 0 | 1 | 1 | 0 | 1 | 4.2 |
| 7 | 3 | 0 | 0 | 1 | 0 | 1 | 1 | 3.1 | 7 | 7 | 1 | 1 | 0 | 1 | 1 | 1 | 6.3 |
| 8 | 2 | 0 | 1 | 1 | 0 | 0 | 0 | 4.0 | 8 | 1 | 1 | 0 | 0 | 1 | 0 | 0 | 3.7 |
| 8 | 3 | 0 | 0 | 0 | 0 | 1 | 0 | 4.6 | 8 | 4 | 1 | 1 | 1 | 1 | 1 | 0 | 4.2 |
| 8 | 8 | 1 | 1 | 0 | 0 | 0 | 1 | 6.4 | 8 | 7 | 0 | 0 | 1 | 1 | 0 | 1 | 4.0 |
| 8 | 6 | 1 | 0 | 1 | 0 | 1 | 1 | 3.5 | 8 | 5 | 0 | 1 | 0 | 1 | 1 | 1 | 5.7 |

The resulting analysis of variance table is shown in Table 8.9.

From the analysis in Table 8.9, the effects which are significant at the 1% level are: Day, $MT$, $MS$, $FT$, $AV$, and $FT*AV$. Additional effects which are significant at the 5% level are: $ET$, $MT*MS$, $FT*ET$, and $MT*FT*PT$. The means which correspond to the significant interaction effects are shown in Table 8.10.

## Table 8.9   An ANOVA for the Data in Table 8.8

| Source | DF | Sum of squares | | Mean square | F value |
|---|---|---|---|---|---|
| Model | 44 | 148.19937500 | | 3.36816761 | 8.41 |
| Error | 19 | 7.61046875 | | 0.40055099 | PR > F |
| Corrected | | | | | |
| Total | 63 | 155.80984375 | | | 0.0001 |

| R-square | C.V. | Root MSE | | Quality mean | |
|---|---|---|---|---|---|
| 0.951155 | 11.5038 | 0.63289098 | | 5.50156250 | |

| Source | DF | ANOVA SS | F value | PR > F |
|---|---|---|---|---|
| Day | 7 | 41.44859375 | 14.78 | 0.0001 |
| MT | 1 | 19.03140625 | 47.51 | 0.0001 |
| MS | 1 | 8.77640625 | 21.91 | 0.0002 |
| FT | 1 | 52.74390625 | 131.68 | 0.0001 |
| AV | 1 | 3.01890625 | 7.54 | 0.0129 |
| ET | 1 | 1.65765625 | 4.14 | 0.0561 |
| PT | 1 | 0.83265625 | 2.08 | 0.1656 |
| MT*MS | 1 | 1.78890625 | 4.47 | 0.0480 |
| MT*FT | 1 | 0.17015625 | 0.42 | 0.5224 |
| MT*AV | 1 | 0.04515625 | 0.11 | 0.7407 |
| MT*ET | 1 | 0.11390625 | 0.28 | 0.6000 |
| MT*PT | 1 | 0.28890625 | 0.72 | 0.4063 |
| MS*FT | 1 | 0.00015625 | 0.00 | 0.9844 |
| MS*AV | 1 | 0.09765625 | 0.24 | 0.6271 |
| MS*ET | 1 | 0.13140625 | 0.33 | 0.5735 |
| MS*PT | 1 | 0.43890625 | 1.10 | 0.3083 |
| FT*AV | 1 | 5.34765625 | 13.35 | 0.0017 |
| FT*ET | 1 | 2.36390625 | 5.90 | 0.0252 |
| FT*PT | 1 | 0.00140625 | 0.00 | 0.9534 |
| AV*ET | 1 | 0.28890625 | 0.72 | 0.4063 |
| AV*PT | 1 | 0.28890625 | 0.72 | 0.4063 |
| ET*PT | 1 | 1.35140625 | 3.37 | 0.0819 |
| MT*MS*FT | 1 | 0.03515625 | 0.09 | 0.7702 |
| MT*MS*AV | 1 | 0.50765625 | 1.27 | 0.2743 |
| MT*MS*ET | 1 | 1.59390625 | 3.98 | 0.0606 |
| MT*FT*ET | 1 | 0.50765625 | 1.27 | 0.2743 |
| MT*FT*PT | 1 | 2.52015625 | 6.29 | 0.0214 |
| MT*AV*ET | 1 | 0.78765625 | 1.97 | 0.1770 |
| MT*AV*PT | 1 | 0.00015625 | 0.00 | 0.9844 |
| MT*ET*PT | 1 | 0.17015625 | 0.42 | 0.5224 |
| MS*FT*AV | 1 | 0.00140625 | 0.00 | 0.9534 |
| MS*FT*ET | 1 | 0.62015625 | 1.55 | 0.2285 |
| MS*FT*PT | 1 | 0.62015625 | 1.55 | 0.2285 |
| MS*AV*PT | 1 | 0.47265625 | 1.18 | 0.2909 |
| MS*ET*PT | 1 | 0.09765625 | 0.24 | 0.6271 |
| FT*AV*ET | 1 | 0.01265625 | 0.03 | 0.8608 |
| FT*AV*PT | 1 | 0.01265625 | 0.03 | 0.8608 |
| AV*ET*PT | 1 | 0.01265625 | 0.03 | 0.8608 |

## Table 8.10  Means for the Data in Table 8.7

MT*FT*PT MEANS†

| MT | FT | PT | N | Q |
|----|----|----|----|----|
| 0 | 0 | 0 | 8 | 5.43750000 |
| 0 | 0 | 1 | 8 | 6.18750000 |
| 0 | 1 | 0 | 8 | 4.11250000 |
| 0 | 1 | 1 | 8 | 4.08750000 |
| 1 | 0 | 0 | 8 | 7.16250000 |
| 1 | 0 | 1 | 8 | 6.85000000 |
| 1 | 1 | 0 | 8 | 4.83750000 |
| 1 | 1 | 1 | 8 | 5.33750000 |

FT*AV MEANS‡

| FT | AV | N | Q |
|----|----|----|----|
| 0 | 0 | 16 | 6.48125000 |
| 0 | 1 | 16 | 6.33750000 |
| 1 | 0 | 16 | 4.08750000 |
| 1 | 1 | 16 | 5.10000000 |

FT*ET MEANS†

| FT | ET | N | Q |
|----|----|----|----|
| 0 | 0 | 16 | 6.76250000 |
| 0 | 1 | 16 | 6.05625000 |
| 1 | 0 | 16 | 4.56250000 |
| 1 | 1 | 16 | 4.62500000 |

MT*MS MEANS

| MT | MS | N | Q |
|----|----|----|----|
| 0 | 0 | 16 | 4.41875000 |
| 0 | 1 | 16 | 5.49375000 |
| 1 | 0 | 16 | 5.84375000 |
| 1 | 1 | 16 | 6.25000000 |

†LSD = .662.
‡LSD = .468.

**Table 8.11   Single-Degree-of-Freedom Sums of Squares for a Full
Interaction Model on the Data in Table 8.7**

Dependent Variable: Q

| Source | DF | Sum of squares | | Mean square | F value |
|---|---|---|---|---|---|
| Model | 63 | 155.80984375 | | 2.47317212 | 99999.99 |
| Error | 0 | 0.00000000 | | 0.00000000 | PR > F |
| Corrected Total | 63 | 155.80984375 | | | 0.0000 |

| R-square | C.V. | Root MSE | | Quality mean | |
|---|---|---|---|---|---|
| 1.000000 | 0.0000 | 0.00000000 | | 5.50156250 | |

| Source | DF | ANOVA SS | F value | PR > F |
|---|---|---|---|---|
| MT | 1 | 19.03140625 | * | * |
| MS | 1 | 8.77640625 | * | * |
| MT*MS | 1 | 1.78890625 | * | * |
| FT | 1 | 52.74390625 | * | * |
| MT*FT | 1 | 0.17015625 | * | * |
| MS*FT | 1 | 0.00015625 | * | * |
| MT*MS * FT | 1 | 0.03515625 | * | * |
| AV | 1 | 3.01890625 | * | * |
| MT*AV | 1 | 0.04515625 | * | * |
| MS*AV | 1 | 0.09765625 | * | * |
| MT*MS*AV | 1 | 0.50765625 | * | * |
| FT*AV | 1 | 5.34765625 | * | * |
| MT*FT*AV | 1 | 1.85640625 | * | * |
| MS*FT*AV | 1 | 0.00140625 | * | * |
| MT*MS*FT*AV | 1 | 0.31640625 | * | * |
| ET | 1 | 1.65765625 | * | * |
| MT*ET | 1 | 0.11390625 | * | * |
| MS*ET | 1 | 0.13140625 | * | * |
| MT*MS*ET | 1 | 1.59390625 | * | * |
| FT*ET | 1 | 2.36390625 | * | * |
| MT*FT*ET | 1 | 0.50765625 | * | * |
| MS*FT*ET | 1 | 0.62015625 | * | * |
| MT*MS*FT*ET | 1 | 0.26265625 | * | * |
| AV*ET | 1 | 0.28890625 | * | * |
| MT*AV*ET | 1 | 0.78765625 | * | * |
| MS*AV*ET | 1 | 8.33765625 | * | * |
| MT*MS*AV*ET | 1 | 1.47015625 | * | * |
| FT*AV*ET | 1 | 0.01265625 | * | * |
| MT*FT*AV*ET | 1 | 0.19140625 | * | * |
| MS*FT*AV*ET | 1 | 0.06890625 | * | * |
| MT*MS*FT*AV*ET | 1 | 0.43890625 | * | * |
| PT | 1 | 0.83265625 | * | * |
| MT*PT | 1 | 0.28890625 | * | * |
| MS*PT | 1 | 0.43890625 | * | * |
| MT*MS*PT | 1 | 5.46390625 | * | * |
| FT*PT | 1 | 0.00140625 | * | * |
| MT*FT*PT | 1 | 2.52015625 | * | * |
| MS*FT*PT | 1 | 0.62015625 | * | * |

**Table 8.11 (Continued)**

| | | | | |
|---|---|---|---|---|
| MT*MS*FT*PT | 1 | 0.03515625 | * | * |
| AV*PT | 1 | 0.28890625 | * | * |
| MT*AV*PT | 1 | 0.00015625 | * | * |
| MS*AV*PT | 1 | 0.47265625 | * | * |
| MT*MS*AV*PT | 1 | 0.62015625 | * | * |
| FT*AV*PT | 1 | 0.01265625 | * | * |
| MT*FT*AV*PT | 1 | 0.01890625 | * | * |
| MS*FT*AV*PT | 1 | 10.64390625 | * | * |
| MT*MS*FT*AV*PT | 1 | 0.43890625 | * | * |
| ET*PT | 1 | 1.35140625 | * | * |
| MT*ET*PT | 1 | 0.17015625 | * | * |
| MS*ET*PT | 1 | 0.09765625 | * | * |
| MT*MS*ET*PT | 1 | 0.17015625 | * | * |
| FT*ET*PT | 1 | 1.47015625 | * | * |
| MT*FT*ET*PT | 1 | 0.11390625 | * | * |
| MS*FT*ET*PT | 1 | 0.58140625 | * | * |
| MT*MS*FT*ET*PT | 1 | 0.00140625 | * | * |
| AV*ET*PT | 1 | 0.01265625 | * | * |
| MT*AV*ET*PT | 1 | 13.41390625 | * | * |
| MS*AV*ET*PT | 1 | 0.28890625 | * | * |
| MT*MS*AV*ET*PT | 1 | 0.40640625 | * | * |
| FT*AV*ET*PT | 1 | 1.12890625 | * | * |
| MT*FT*AV*ET*PT | 1 | 0.47265625 | * | * |
| MS*FT*AV*ET*PT | 1 | 0.37515625 | * | * |
| MT*MS*FT*AV*ET*PT | 1 | 0.47265625 | * | * |

Examination of the $MT * FT * PT$ means in Table 8.10 reveals that in order to get a high-quality product, one must have $FT$ at a low level and $MT$ at a high level, and that for this combination, the value of $PT$ has little effect. Examination of the $FT * AV$ means in Table 8.10 reveals that in order to get a high-quality product, one must have $FT$ at a low level, and when this is the case, the value of $AV$ has little effect. Examination of the $FT * ET$ means in Table 8.10 reveals that the highest-quality product results when $FT$ is at a low level and when $ET$ is at a low level. Examination of the $MT * MS$ means in Table 8.10 reveals that the highest-quality product should result when both $MT$ and $MS$ are at high levels. None of the interactions of other factors with $PT$ or the main effect of $PT$ is significant, hence it makes little difference which value of $PT$ is chosen. Summarizing, to make a product which has the highest possible quality one should use a high mixing time, a high mixing speed, a low freezing temperature, and a low end product temperature, and for this combination of factors it makes no difference whether one uses a high or low value of air velocity. In addition, the package type has little affect on product quality.

The data in Table 8.8 were also analyzed using the half-normal plotting method. The data were first analyzed by using the following SAS® commands:

```
PROC ANOVA;
CLASSES MT MS FT AV ET PT;
MODEL Q = MT|MS|FT|AV|ET|PT;
```

The analysis of variance table obtained from this analysis is shown in Table 8.11. A half-normal plot of the effects in Table 8.11 is shown in Figure 8.1.

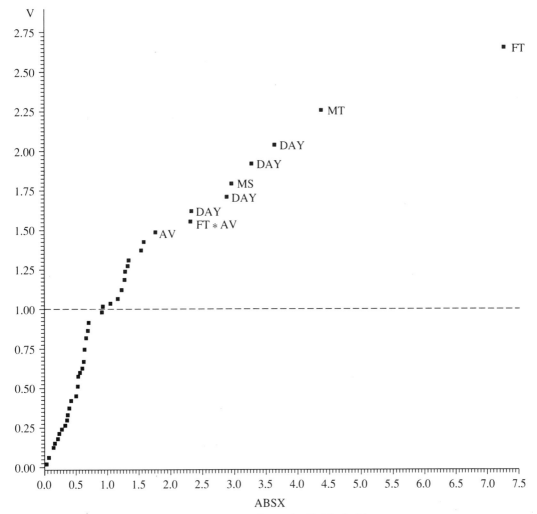

**Figure 8.1   A Half-Normal Plot of the Effects in Table 8.11.**

Examination of the plot in Figure 8.1 reveals that the factors which are most significant are $FT$, $MT$, $MS$, and $FT * AV$. Four of the interaction contrasts which were confounded with Days are also significant. These are labeled with Day in Figure 8.1. A half-normal plot was also constructed from the effects in Table 8.10 after removing the eight significant effects identified from the plot in Figure 8.1. This plot is shown in Figure 8.2. Examination of this plot would lead one to believe that there are no other significant effects and an estimate of $\sigma$ is $\hat{\sigma} = .74$. Using this estimate of $\sigma$, other effects in Table 8.11 which have single-degree-of-freedom sums of squares values larger than $(2\hat{\sigma})^2 = (1.48)^2 = 2.19$ are $AV$, $MT * FT * PT$, and $FT * ET$. These results are consistent

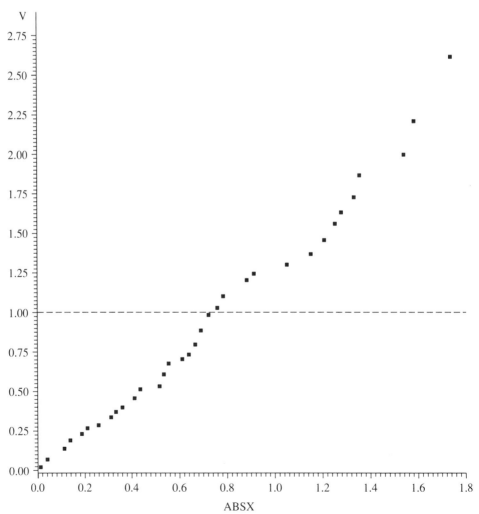

**Figure 8.2    A Half-Normal Plot of the Effects in Table 8.10 After Removing the Eight Largest Effects.**

with those found in the previous analysis. The only difference between the two is that $MT * MS$ was significant in the first analysis but does not appear to be significant in this analysis.

In the next chapter, the design and analysis of factorial experiments without obtaining a full replication of the treatment combinations will be considered.

# 9

# Fractional Replications of $2^n$ Factorial Treatment Structures

## CHAPTER OUTLINE

I n Chapter 8 it was noted that high-order interaction effects are rarely significant. Because of this it was shown that one is able to conduct experiments more efficiently by confounding some of the high-order interaction effects with blocking effects. As another consequence of this observation, it is often possible to obtain information about all main effects and low-order interaction effects without even completing a full replication of the factorial experiment, a tremendous advantage for experiments involving large numbers of factors. The only disadvantage in conducting less than a full replicate of a factorial experiment is that many of the main effects and low-order interaction effects will be confounded with high-order interaction effects. But if the higher-order interaction effects rarely exist, then no loss of information about the low-order effects occurs. The design and analysis of these kinds of experiments is the subject of this chapter.

To further illustrate, consider a factorial experiment involving seven factors with each factor at two levels. Such an experiment would require 128 trials for a full replicate. In a $2^7$ experiment, there are 7 main effects, 21 two-factor interaction effects, and 35 three-factor interaction effects, giving a total of 63 low-order effects. If one were to conduct a full run of this experiment, there would be 64 degrees of freedom for estimating the experimental error variance. If chosen properly, one is able to perform only 64 trials, one-half of a full run, and still estimate all of the important effects as well as the experimental error variance as three-factor interactions generally tend to be much less significant than main effects and two-factor interactions. There are 28 two-factor interactions and main effects. With an appropriate selection of experimental runs, all of these 28 effects could be estimated with 32 trials and still one should be able to estimate the experimental error variance. In a screening experiment designed to select those factors most likely to have the most impact on the response variables of interest, it might be possible to determine those factors with as few as 16 trials. Obviously, knowledge and use of these so-called fractional factorial experiments can have a significant impact on the productivity of researchers.

## 9.1 A ONE-HALF REPLICATE OF A $2^3$ FACTORIAL EXPERIMENT

The basic ideas required for understanding fractional factorial experiments will be illustrated by considering a one-half replicate of a $2^3$ factorial experiment even though such a small experiment could not in general be recommended.

Determining an appropriate set of treatment combinations to run is extremely easy using the ideas suggested in Chapter 8. To illustrate, in Chapter 8, methods were introduced which would enable one to conduct a $2^3$ experiment in two blocks by confounding the three-factor interaction with blocks. To conduct a one-half replicate of a $2^3$ experiment, one would simply select one of these two sets of four treatment combinations at random and run the trials suggested by that half. Table 8.1 shows the division of the set of eight treatment combinations in a $2^3$ experiment into two groups by confounding the three-factor interaction. One would simply select one of these sets by some random procedure. Suppose one selects the second set. Then an experiment would be conducted using the four treatment combinations in this set, namely, $(0,0,0)$, $(1,1,0)$, $(1,0,1)$, and $(0,1,1)$ or $(1)$, $ab$, $ac$, and $bc$. The order in which these four experiments would be conducted would be selected at random.

Table 9.1 shows the columns out of Table 7.2 that correspond to the treatment combinations selected above. These columns illustrate the linear combinations of the observed values of these treatment combinations which define each of the effects in a $2^3$ experiment.

Notice that in Table 9.1, row 8 is the opposite of row 1, row 7 is the opposite of row 2, row 6 is the opposite of row 3, and row 4 is the opposite of row 5. Thus the linear combination

$$(1) + (ab) + (ac) + (bc)$$

estimates the Total $- A*B*C$ effect, while the contrast

$$-(1) + (ab) + (ac) - (bc)$$

estimates $A - B*C$. The contrast

$$-(1) + (ab) - (ac) + (bc)$$

estimates $B - A*C$, and the contrast

$$-(1) - (ab) + (ac) + (bc)$$

estimates $C - A*B$.

**Table 9.1   Coefficients for the Effects in a One-Half Replicate of a $2^3$ Factorial Experiment**

| | TREATMENT COMBINATION | | | |
| --- | --- | --- | --- | --- |
| Effect | (1) | ab | ac | bc |
| Total | + | + | + | + |
| A | − | + | + | − |
| B | − | + | − | + |
| A * B | + | + | − | − |
| C | − | − | + | + |
| A * C | + | − | + | − |
| B * C | + | − | − | + |
| A * B * C | − | − | − | − |

If the other half of the treatment combinations were run, the linear combination

$$(a) + (b) + (c) + (abc)$$

would estimate Total $+ A*B*C$, while the contrast

$$(a) - (b) - (c) + (abc)$$

would estimate $A + B*C$. The contrast

$$-(a) + (b) - (c) + (abc)$$

would estimate $B + A*C$, and the contrast

$$-(a) - (b) + (c) + (abc)$$

would estimate $C + B*C$. In either case note that $A$ is confounded with $B*C$, $B$ is confounded with $A*C$, $C$ is confounded with $A*B$, and $A*B*C$ is confounded with the overall total effect. It is customary to call two effects which are estimated from the same or opposite linear combinations of the treatment combinations *aliases*. Thus $A$ and $B*C$ are aliases as are $B$ and $A*C$, and $C$ and $A*B$.

A quicker and much easier way to determine the alias sets is to determine the generalized interaction, as defined in Chapter 8, between an effect and the defining contrast. For example, the generalized interaction between $A$ and $A*B*C$ is

$$(A)*(A*B*C) = B*C$$

since $A$ occurs twice and hence cancels out. Therefore $A$ is aliased with $B*C$. In a similar manner, the alias of $B$ is $A*C$ since the generalized interaction between $B$ and $A*B*C$ is

$$(B)*(A*B*C) = A*C$$

And the alias of $C$ is $A*B$ since the generalized interaction between $C$ and $A*B*C$ is

$$(C)*(A*B*C) = A*B$$

If one were to analyze a one-half replicate of the above $2^3$ experiment and compute sums of squares corresponding to each of the effects in a $2^3$ experiment, one would find that the sum of squares for $A$ would be identical to the sum of squares for $B*C$. Thus if the sum of squares for $A$ and/or $B*C$ is large, one would not be able to tell whether the largeness of this sum of squares is due to $A$ or $B*C$ or possibly both. Even if this sum of squares were small, one could not say with assurance that neither effect is present, since it is possible that the two effects could be canceling out. Thus it is very important that the low-order effects be aliases of high-order interaction effects only as these effects are not likely to be real. Similar results would hold for the other alias sets.

In the next section, a one-half replicate of a $2^5$ factorial experiment is considered.

**9.2**
**A ONE-HALF**
**REPLICATE**
**OF A $2^5$**
**FACTORIAL**
**EXPERIMENT**

To determine the 16 runs to make in a one-half replicate of a $2^5$ factorial experiment involving factors $A$, $B$, $C$, $D$, and $E$, one could select those treatment combinations which satisfy:

$$\ell(A) + \ell(B) + \ell(C) + \ell(D) + \ell(E) = 0$$

or those that satisfy

$$\ell(A) + \ell(B) + \ell(C) + \ell(D) + \ell(E) = 1$$

as these are the two conditions which would confound the $A*B*C*D*E$ interaction with blocks if one were blocking a $2^5$ factorial experiment in two blocks. Suppose by some random procedure, the combinations satisfying the second equation above were selected; then the treatment combinations which would be run are as given in Table 9.2.

Table 9.3 shows the alias sets for this set of treatment combinations. Note that each main effect has a four-factor interaction as an alias and

**Table 9.2   A Set of Treatment Combinations for a
One-Half Replicate of a $2^5$ Factorial Experiment**

| | TREATMENT COMBINATION | | | | |
|---|---|---|---|---|---|
| *Run* | *A* | *B* | *C* | *D* | *E* |
| 12 | 1 | 0 | 0 | 0 | 0 |
| 8 | 0 | 1 | 0 | 0 | 0 |
| 2 | 0 | 0 | 1 | 0 | 0 |
| 15 | 0 | 0 | 0 | 1 | 0 |
| 7 | 0 | 0 | 0 | 0 | 1 |
| 9 | 1 | 1 | 1 | 0 | 0 |
| 11 | 1 | 1 | 0 | 1 | 0 |
| 14 | 1 | 1 | 0 | 0 | 1 |
| 1 | 1 | 0 | 1 | 1 | 0 |
| 5 | 1 | 0 | 1 | 0 | 1 |
| 13 | 1 | 0 | 0 | 1 | 1 |
| 4 | 0 | 1 | 1 | 1 | 0 |
| 10 | 0 | 1 | 1 | 0 | 1 |
| 16 | 0 | 1 | 0 | 1 | 1 |
| 6 | 0 | 0 | 1 | 1 | 1 |
| 3 | 1 | 1 | 1 | 1 | 1 |

**Table 9.3   Alias Sets for a One-Half Replicate of a $2^5$ Factorial Experiment Defined by the $A*B*C*D*E$ Interaction**

| Effect | Alias |
|--------|-------|
| $A$ | $B*C*D*E$ |
| $B$ | $A*C*D*E$ |
| $C$ | $A*B*D*E$ |
| $D$ | $A*B*C*E$ |
| $E$ | $A*B*C*D$ |
| $A*B$ | $C*D*E$ |
| $A*C$ | $B*D*E$ |
| $A*D$ | $B*C*E$ |
| $A*E$ | $B*C*D$ |
| $B*C$ | $A*D*E$ |
| $B*D$ | $A*C*E$ |
| $B*E$ | $A*C*D$ |
| $C*D$ | $A*B*E$ |
| $C*E$ | $A*B*D$ |
| $D*E$ | $A*B*C$ |

each two-factor interaction has a three-factor interaction as an alias. The quantities which could be estimated from data collected from the set of treatment combinations in Table 9.2 are the Effect ± the Alias for each (Effect,Alias) pair in Table 9.3.

## 9.3  THE ANALYSIS OF FRACTIONAL FACTORIAL EXPERIMENTS

Both the half-normal plot method and the analysis of variance method discussed in Section 7.4 could be used to help analyze fractional factorial experiments. Which is more appropriate depends on whether there are effects and their aliases which can be assumed to be measuring experimental error. This in turn depends on the size of the experiment and effects occurring in the alias sets. In a one-half replicate of a $2^5$ factorial experiment, one would not generally be willing to assume that the two-factor interactions could be pooled to get an estimate of the experimental error variance. Thus the appropriate method for analyzing such an experiment would be to use the half-normal plotting method. However, in a one-half replicate of a $2^6$ factorial experiment with treatment combinations selected by using the six-factor interaction as the defining contrast, there are six main effects, each aliased with a five-factor interaction and 15 two-factor interactions each aliased with a four-factor interaction. Each three-factor interaction has another three-factor interaction as its alias pair, giving 10 alias pairs of three-factor interactions. If these were believed to be estimating experimental error, the sum of squares corresponding to each one of these could be pooled to give an estimate of $\sigma^2$ based on 10 degrees of freedom. If one believed this assumption might be suspect, one could still construct a half-normal plot of the 31 effects defined by the alias pairs.

The easiest way to get the information needed for a half-normal plot of the effects corresponding to the alias sets in a fractional factorial experiment is to use an ANOVA procedure, fitting a model which includes a term for one member of each alias set. That is, for the experiment described in Section 9.2 and using SAS® ANOVA, one would use the statements:

PROC ANOVA;
CLASSES A B C D E;
MODEL dep. vars. = A B C D E A∗B A∗C A∗D
    A∗E B∗C B∗D B∗E C∗D C∗E D∗E;

This analysis will give 15 single-degree-of-freedom sums of squares each corresponding to one alias pair. The half-normal plotting technique would be applied to the square roots of these single-degree-of-freedom sums of squares for each of the 15 effects in Table 9.3.

**Table 9.4   Results from a One-Half
Replicate of a $2^5$ Factorial Experiment**

| | | TREATMENT COMBINATION | | | |
|---|---|---|---|---|---|
| W | M | T | C | P | Quality |
| 0 | 0 | 0 | 0 | 0 | 4.8 |
| 1 | 1 | 0 | 0 | 0 | 2.2 |
| 1 | 0 | 1 | 0 | 0 | 4.2 |
| 0 | 1 | 1 | 0 | 0 | 3.0 |
| 1 | 0 | 0 | 1 | 0 | 2.2 |
| 0 | 1 | 0 | 1 | 0 | 8.4 |
| 0 | 0 | 1 | 1 | 0 | 5.3 |
| 1 | 1 | 1 | 1 | 0 | 8.9 |
| 1 | 0 | 0 | 0 | 1 | 5.0 |
| 0 | 1 | 0 | 0 | 1 | 5.8 |
| 0 | 0 | 1 | 0 | 1 | 4.6 |
| 1 | 1 | 1 | 0 | 1 | 5.2 |
| 0 | 0 | 0 | 1 | 1 | 2.9 |
| 1 | 1 | 0 | 1 | 1 | 6.6 |
| 1 | 0 | 1 | 1 | 1 | 2.7 |
| 0 | 1 | 1 | 1 | 1 | 7.0 |

To illustrate, consider the example described in Section 7.4. Suppose the experimenter only ran those treatment combinations which satisfied

$$\ell(W) + \ell(M) + \ell(T) + \ell(C) + \ell(P) = 0$$

Hence the resulting data would appear as in Table 9.4.

The data in Table 9.4 were analyzed using SAS® ANOVA with the following control statements:

PROC ANOVA;
CLASSES W M T C P;
MODEL QUALITY = W M W∗M T W∗T M∗T C W∗C M∗C
T∗C P W∗P M∗P T∗P C∗P;

This analysis produced the results in Table 9.5. A half-normal plot of the effects in Table 9.5 is shown in Figure 9.1.

### Table 9.5 An ANOVA for the Data in Table 9.4 Fitting All Main Effects and Two-Factor Interactions

*Dependent Variable: Quality*

| Source | DF | Sum of squares | | Mean square | F value |
|---|---|---|---|---|---|
| Model | 15 | 64.23000000 | | 4.28200000 | 99999.99 |
| Error | 0 | 0.00000000 | | 0.00000000 | PR > F |
| Corrected Total | 15 | 64.23000000 | | | 0.0000 |

| R-square | C.V. | Root MSE | | Quality mean | |
|---|---|---|---|---|---|
| 1.000000 | 0.0000 | 0.00000000 | | 4.92500000 | |

| Source | DF | ANOVA SS | F value | PR > F |
|---|---|---|---|---|
| W | 1 | 1.44000000 | * | * |
| M | 1 | 14.82250000 | * | * |
| W∗M | 1 | 0.30250000 | * | * |
| T | 1 | 0.56250000 | * | * |
| W∗T | 1 | 3.06250000 | * | * |
| M∗T | 1 | 0.04000000 | * | * |
| C | 1 | 5.29000000 | * | * |
| W∗C | 1 | 0.16000000 | * | * |
| M∗C | 1 | 25.50250000 | * | * |
| T∗C | 1 | 1.32250000 | * | * |
| P | 1 | 0.04000000 | * | * |
| W∗P | 1 | 0.64000000 | * | * |
| M∗P | 1 | 0.72250000 | * | * |
| T∗P | 1 | 1.32250000 | * | * |
| C∗P | 1 | 9.00000000 | * | * |

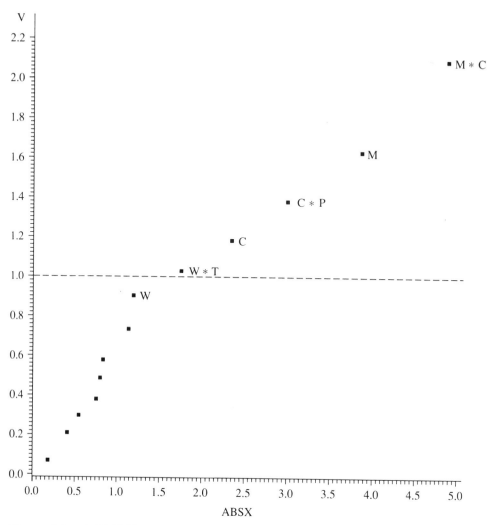

**Figure 9.1   A Half-Normal Plot of the Effects in Table 9.5.**

Examination of the plot in Figure 9.1 reveals that the effects which appear to be significant are $M*C$, $M$, $C*P$, $C$, and $W*T$. These five effects were removed and a half-normal plot of the remaining 10 effects was obtained. This plot is shown in Figure 9.2. From this plot it would appear that an estimate of $\sigma$ is $\hat{\sigma} = .83$. The analysis in Chapter 7 also showed that $M*C$, $M$, $C*P$, and $C$ were significant. The only major differences in the two analyses are that (1) the Chapter 7 analysis showed that $W$ was significant while in this analysis $W$ does not appear to be significant (it is interesting to note that $W$ is the next largest effect), (2) the Chapter 7 analysis showed that the $M*C*P$ effect was significant while this analysis shows that the $W*T$ effect is significant (note that these

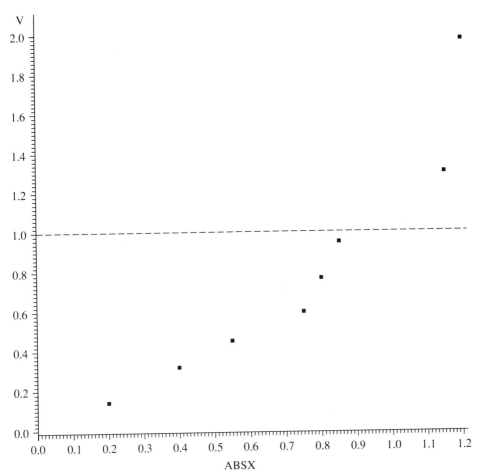

**Figure 9.2   A Half-Normal Plot of the Effects in Table 9.5 After Removing the Five Largest Effects.**

two effects are alias pairs in this example), (3) the Chapter 7 analysis showed that the $W*M*C$ effect was significant (the alias of this three-factor interaction is $T*P$, but it is not significant here—it is the next largest effect after the $W$ effect), and (4) the estimate of $\sigma$ was .61 in the Chapter 7 analysis and is .83 in this analysis.

**9.4 SMALLER FRACTIONS OF $2^n$ FACTORIAL EXPERIMENTS**

In this section, smaller fractions of the $2^n$ factorial experiments will be considered. To illustrate, consider a $2^6$ factorial experiment with factors $A$, $B$, $C$, $D$, $E$, and $F$. A full replicate would require 64 experimental runs, while a half-replicate would require 32 runs. What could be learned from a one-fourth replicate which would require 16 runs? What can be learned depends upon how one selects the 16 treatment factor combinations to run.   Obviously one would want to select treatment combinations using high-order interaction contrasts as the defining contrasts. It should also be clear that rather than having two effects in each alias set, that there will be four effects in each alias set.

Suppose one selects those combinations which satisfy:

$$L_1 = \ell(A) + \ell(B) + \ell(C) + \ell(D) + \ell(E) + \ell(F) = \text{odd}$$

$$L_2 = \ell(A) + \ell(B) + \ell(C) = \text{even} \qquad (9.4.1)$$

Then

$$L_1 = \ell(D) + \ell(E) + \ell(F) = \text{odd}$$

Each effect will have three aliases, and these alias sets can be determined from the generalized interactions between the given effect and each of the defining contrasts.  For example, the aliases of the main effect of factor $A$ are

$$(A)*(A*B*C*D*E*F) = B*C*D*E*F$$

$$(A)*(A*B*C) = B*C$$

and

$$(A)*(D*E*F) = A*D*E*F$$

**Table 9.6    Alias Sets Resulting from Using the**
**$A*B*C*D*E*F$ and $A*B*C$ Effects as Defining Contrasts**

| Effect | Alias 1 | Alias 2 | Alias 3 |
|--------|---------|---------|---------|
| (1) | $A*B*C*D*E*F$ | $A*B*C$ | $D*E*F$ |
| $A$ | $B*C*D*E*F$ | $B*C$ | $A*D*E*F$ |
| $B$ | $A*C*D*E*F$ | $A*C$ | $B*D*E*F$ |
| $C$ | $A*B*D*E*F$ | $A*B$ | $C*D*E*F$ |
| $D$ | $A*B*C*E*F$ | $A*B*C*D$ | $E*F$ |
| $E$ | $A*B*C*D*F$ | $A*B*C*E$ | $D*F$ |
| $F$ | $A*B*C*D*E$ | $A*B*C*F$ | $D*E$ |
| $A*D$ | $B*C*E*F$ | $B*C*D$ | $A*E*F$ |
| $A*E$ | $B*C*D*F$ | $B*C*E$ | $A*D*F$ |
| $A*F$ | $B*C*D*E$ | $B*C*F$ | $A*D*E$ |
| $B*D$ | $A*C*E*F$ | $A*C*D$ | $B*E*F$ |
| $B*E$ | $A*C*D*F$ | $A*C*E$ | $B*D*F$ |
| $B*F$ | $A*C*D*E$ | $A*C*F$ | $B*D*E$ |
| $C*D$ | $A*B*E*F$ | $A*B*D$ | $C*E*F$ |
| $C*E$ | $A*B*D*F$ | $A*B*E$ | $C*D*F$ |
| $C*F$ | $A*B*D*E$ | $A*B*F$ | $C*D*E$ |

If one were to run the treatment combinations defined by (9.4.1), then the contrast corresponding to the $A$ effect would be estimating

$$A \pm B*C*D*E*F \pm B*C \pm A*D*E*F$$

and would estimate $A$ only if $B*C*D*E*F$, $B*C$, and $A*D*E*F$ were truly equal to zero.

The aliases of $A*B$ are

$$(A*B)*(A*B*C*D*E*F) = C*D*E*F$$

$$(A*B)*(A*B*C) = C$$

and

$$(A*B)*(D*E*F) = A*B*D*E*F$$

Table 9.6 shows the alias sets resulting from the defining contrasts in (9.4.1).

The biggest problem with the above design is that all of the main effects have two-factor interactions as aliases. Is it possible to construct a one-fourth replicate of a $2^6$ factorial experiment and not have any main effects or two-factor interactions confounded with other main effects or two-factor interactions? There are 6 main effects to be estimated and 15 two-factor interactions to be estimated, giving a total of 21 low-order effects to be estimated. Obviously, it is not possible to estimate all of these with the 16 observations in a one-fourth replicate. It is

### Table 9.7 Alias Sets Resulting from Using the $A*B*C*D$ and $A*B*E*F$ Effects as Defining Contrasts

| Effect | Alias 1 | Alias 2 | Alias 3 |
|---|---|---|---|
| (1) | $A*B*C*D$ | $A*B*E*F$ | $C*D*E*F$ |
| $A$ | $B*C*D$ | $B*E*F$ | $A*C*D*E*F$ |
| $B$ | $A*C*D$ | $A*E*F$ | $B*C*D*E*F$ |
| $C$ | $A*B*D$ | $A*B*C*E*F$ | $D*E*F$ |
| $D$ | $A*B*C$ | $A*B*D*E*F$ | $C*E*F$ |
| $E$ | $A*B*C*D*E$ | $A*B*F$ | $C*D*F$ |
| $F$ | $A*B*C*D*F$ | $A*B*E$ | $C*D*E$ |
| $A*B$ | $C*D$ | $E*F$ | $A*B*C*D*E*F$ |
| $A*C$ | $B*D$ | $B*C*E*F$ | $A*D*E*F$ |
| $A*D$ | $B*C$ | $B*D*E*F$ | $A*C*E*F$ |
| $A*E$ | $B*C*D*E$ | $B*F$ | $A*C*D*F$ |
| $A*F$ | $B*C*D*F$ | $B*E$ | $A*C*D*E$ |
| $C*E$ | $A*B*D*E$ | $A*B*C*F$ | $D*F$ |
| $C*F$ | $A*B*D*F$ | $A*B*C*E$ | $D*E$ |
| $A*C*E$ | $B*D*E$ | $B*C*F$ | $A*D*F$ |
| $A*C*F$ | $B*D*F$ | $B*C*E$ | $A*D*E$ |

possible, however, to design an experiment so that the main effects are not confounded with any two-factor interactions or other main effects, but some of the two-factor interactions will be confounded with other two-factor interactions. To illustrate, let the defining effects be

$$L_1 = \ell(A) + \ell(B) + \ell(C) + \ell(D) = \text{even}$$

and

$$L_2 = \ell(A) + \ell(B) + \ell(E) + \ell(F) = \text{even} \qquad (9.4.2)$$

then

$$L_3 = \ell(C) + \ell(D) + \ell(E) + \ell(F) = \text{even}$$

The alias sets from this set of defining effects are given in Table 9.7.

If an experimenter has some prior information that would tend to make him or her believe that certain of the factors would not interact, then this information could be used to help design the experiment. For example, if it were known that factor $A$ would probably not interact with any other factors, factors $E$ and $F$ would probably not interact, and if it were true that three-factor and higher-order interactions were not real, then all main effects and two-factor interactions could be estimated from the above design except for the $C*F$ and $D*E$ effects, which are in the same alias set.

To analyze data collected from the above experimental design, one would simply use an ANOVA procedure and fit a model which contains

**Table 9.8   Data from a One-Fourth Replicate of a $2^6$ Factorial Experiment with Treatment Combinations Selected by (9.4.2)**

| A | B | C | D | E | F | y |
|---|---|---|---|---|---|---|
| 0 | 0 | 0 | 0 | 0 | 0 | 41 |
| 1 | 1 | 0 | 0 | 0 | 0 | 41 |
| 0 | 0 | 1 | 1 | 0 | 0 | 46 |
| 0 | 0 | 0 | 0 | 1 | 1 | 36 |
| 1 | 1 | 1 | 1 | 1 | 1 | 62 |
| 0 | 0 | 1 | 1 | 1 | 1 | 29 |
| 1 | 1 | 0 | 0 | 1 | 1 | 45 |
| 1 | 1 | 1 | 1 | 0 | 0 | 78 |
| 1 | 0 | 1 | 0 | 1 | 0 | 35 |
| 1 | 0 | 1 | 0 | 0 | 1 | 36 |
| 1 | 0 | 0 | 1 | 1 | 0 | 25 |
| 1 | 0 | 0 | 1 | 0 | 1 | 41 |
| 0 | 1 | 1 | 0 | 1 | 0 | 47 |
| 0 | 1 | 1 | 0 | 0 | 1 | 34 |
| 0 | 1 | 0 | 1 | 1 | 0 | 58 |
| 0 | 1 | 0 | 1 | 0 | 1 | 74 |

terms for one of the effects in each alias set. To analyze the above design using SAS® one could use

PROC ANOVA;
CLASSES A B C D E F;
MODEL A B C D E F A∗B A∗C A∗D A∗E A∗F
    C∗E C∗F A∗C∗E A∗C∗F;                                    (9.4.3)

and then construct a half-normal plot of the square roots of the sums of squares corresponding to each of the effects in the model. One could also remove the two three-factor interaction terms from the above model, after which the sums of squares for these two effects would be pooled to provide a 2-degree-of-freedom estimate of $\sigma^2$. This would not generally be our recommendation, however. Unless one has a few more degrees of freedom for their estimate of experimental error, the estimate can be too sensitive to a few abnormalities in the data. Thus, one would usually be well advised to use the half-normal plotting technique to help analyze fractional factorial experiments.

#### Table 9.9   An Analysis of Variance Table for the Data in Table 9.8

*Dependent Variable: Y*

| Source | DF | Sum of squares | Mean square | F value |
|---|---|---|---|---|
| Model | 15 | 3476.00000000 | 231.73333333 | 99999.99 |
| Error | 0 | 0.00000000 | 0.00000000 | PR > F |
| Corrected Total | 15 | 3476.00000000 | | 0.0000 |

| R-square | C.V. | Root MSE | Y mean |
|---|---|---|---|
| 1.000000 | 0.0000 | 0.00000000 | 45.50000000 |

| Source | DF | ANOVA SS | F value | PR > F |
|---|---|---|---|---|
| A | 1 | 0.25000000 | * | * |
| B | 1 | 1406.25000000 | * | * |
| C | 1 | 2.25000000 | * | * |
| D | 1 | 600.25000000 | * | * |
| E | 1 | 182.25000000 | * | * |
| F | 1 | 12.25000000 | * | * |
| A∗B | 1 | 49.00000000 | * | * |
| A∗C | 1 | 784.00000000 | * | * |
| A∗D | 1 | 0.00000000 | ·* | * |
| A∗E | 1 | 1.00000000 | * | * |
| A∗F | 1 | 36.00000000 | * | * |
| C∗E | 1 | 9.00000000 | * | * |
| C∗F | 1 | 361.00000000 | * | * |
| A∗C∗E | 1 | 30.25000000 | * | * |
| A∗C∗F | 1 | 2.25000000 | * | * |

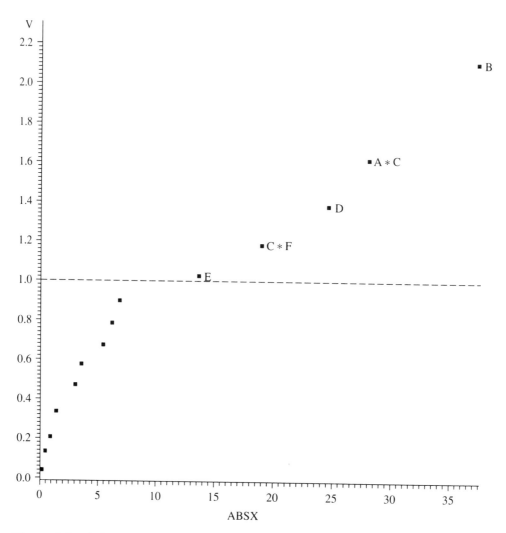

**Figure 9.3   A Half-Normal Plot of the Effects in Table 9.9.**

This section concludes with an example. Suppose an experiment is conducted according to the defining effects in (9.4.2) and gives the data in Table 9.8.

The data in Table 9.8 were first analyzed with SAS® using the model in (9.4.3). The results are shown in Table 9.9. Next a half-normal plot of the effects in Table 9.9 was constructed, and the results are shown in Figure 9.3. Examination of this plot reveals that five effects appear to be significant. These five effects are $B$, $A*C$, $D$, $C*F$, and $E$. Figure 9.4 shows a half-normal plot of the remaining 10 effects. An estimate of $\sigma$ from this plot is $\hat{\sigma} = 3.8$.

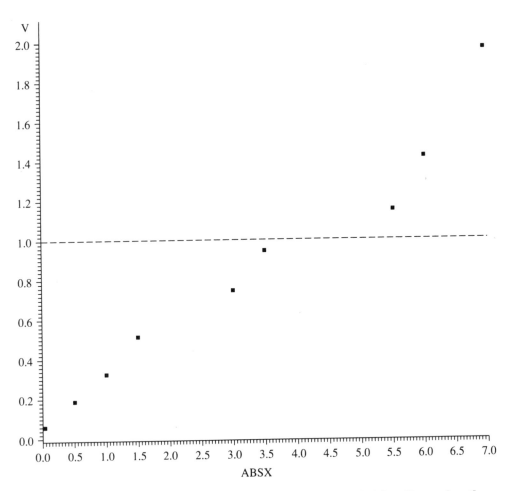

**Figure 9.4   A Half-Normal Plot of the Effects in Table 9.9 After Removing the Five Largest Effects.**

There has also been a significant amount of work done with 3$^n$ factorial experiments. These experiments involve $n$ factors with each factor occurring at three levels—low, medium, and high, usually denoted by 0, 1, and 2, respectively. For experiments involving factors at three levels, the 2 degrees of freedom corresponding to a main effect can be partitioned into a single-degree-of-freedom sum of squares measuring the linear effect of that treatment factor and a single-degree-of-freedom sum of squares measuring the quadratic effect of the treatment factor. Two-factor interaction sums of squares having 4 degrees of freedom each can be partitioned into Linear*Linear, Linear*Quadratic, Quadratic*Linear, and Quadratic*Quadratic effects. Three-factor interaction sums of squares with 8 degrees of freedom each can be partitioned into Linear*Linear

*Linear, Linear*Linear*Quadratic, . . . effects. High-order interaction effects can be confounded with Blocks as was done in Chapter 8 for $2^n$ factorial experiments. For example, a $3^3$ factorial experiment requiring 27 treatment combinations per complete replicate could be conducted in three blocks of size 9 each by confounding the Linear*Quadratic*Quadratic and the Quadratic*Quadratic*Quadratic interactions with blocks. Choosing to conduct the treatment combinations in only one of these blocks would provide a one-third replicate of a $3^3$ factorial experiment from which one could estimate each of the main effects. While such designs are interesting from a theoretical point of view, we have used them very little in practice. If one has quantitative treatment factors and is going to allow more than two levels per treatment factor, then we believe that the response surface methodology discussed in Chapter 11 is much better suited to these situations. For these reasons, the design and analysis of $3^n$ factorial experiments is not considered in this book. The interested reader can see Cochran and Cox (1957), Hicks (1982), or Kempthorne (1952) for further information on this subject.

# 10

# Polynomial Model Approximations

$A$n experiment was conducted in order to determine the effects of storage temperature on potatoes. Six refrigerators were available and six different temperatures, 35°, 38°, 41°, 44°, 47°, and 50°F, were randomly assigned to the refrigerators, one to each refrigerator. Three 10-pound bags of potatoes were placed in each refrigerator. After 25 weeks of storage, four potatoes were randomly selected from each of the three bags in each of the six refrigerators. Several indicators of quality were then measured on each of the sampled potatoes.

The experimental units for this experiment are the refrigerators, and although we have 12 measurements on each refrigerator, these measures do not provide independent replications of the six different temperature levels. In actuality, only one independent piece of information is available for each temperature level, the mean of the 12 observed values for each quality variable. Once again, one is faced with trying to gather useful information from a nonreplicated experiment.

The half-normal plotting techniques could be used on these data, but there are only five orthogonal contrasts for comparing the six temperature levels. With such a few number of orthogonal contrasts, the half-normal technique will generally not be very successful. Another way to try to gain some information from this experiment is to assume the effects of temperature can be modeled by a polynomial model. Analyzing such models is the subject covered in this chapter.

**10.1
A
CAUTIONARY
NOTE**

In those cases where the treatments under study or populations being sampled correspond to different levels of a quantitative factor, it is often possible to obtain good information from collected data by assuming that the true effects, $(\mu_1, \mu_2, \ldots, \mu_t)$, of $t$ factor levels can be modeled by a simple polynomial model such as

$$\mu_i = \beta_0 + \beta_1 X_i + \beta_2 X_i^2 + \cdots + \beta_k X_i^k$$

$$i = 1, 2, \ldots, t$$

where $X_i$ represents the value of the $i$th level of the quantitative treatment factor, $i = 1, 2, \ldots, t$. Such an assumption is not unreasonable, given that the true state of nature does not play tricks on us. In choosing such a model it is important that the degree of the model selected does not approach $t - 1$. It is always possible to fit $t$ points exactly with a $t - 1$ degree polynomial, and although such a model fits the observed data very well, one cannot be at all confident that such a polynomial represents the true state of nature. To illustrate, consider the data plot in Figure 10.1 and the two fitted polynomial functions represented by $A$ and $B$. Even though function $B$, a fifth-degree polynomial goes through each of the data points, it is obvious that it likely does not represent the true state of nature. The straight line given by function $A$ seems to be a much more

**148**

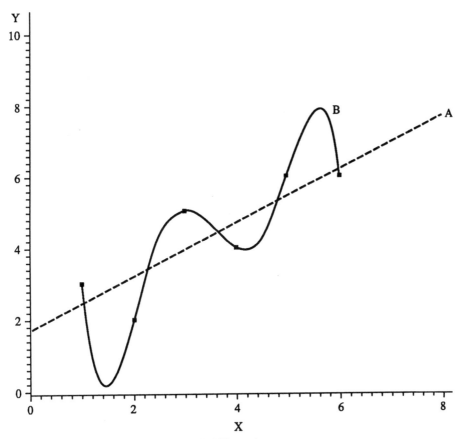

**Figure 10.1   Data and Polynomial Functions.**

reasonable approximation to the true state of nature even though it does not go through all of the data points. When fitting polynomial models to data, it is always a good idea to plot the model not only at those factor levels where you have data but also plot the model in between the points where you have data. Such a plot provides a good visual check as to whether the model selected is a reasonable one or whether one is "overfitting" the data.

It should be noted that experimental situations rarely require polynomial models of degrees larger than 3. Usually, a simple linear model or a quadratic model will be adequate. If none of these lower-degree polynomial models appears to fit the data very well, then one should probably consider using a model which is nonlinear in the parameters (called a nonlinear model) to model the data. It might be noted that polynomial models, while being nonlinear in the factor levels, are linear in the parameters and, hence, are called linear models by statisticians.

**10.2
CHOOSING A
DEGREE**

To decide on the degree required for an adequate polynomial model approximation to the data, one should plot the data to see whether there appears to be a linear or curvilinear trend in the data.

If the trend appears to be linear, one should fit the quadratic model

$$y = \beta_0 + \beta_1 X + \beta_2 X^2 + \epsilon$$

to the data and test $H_{02}$: $\beta_2 = 0$. If $H_{02}$ is rejected, then one should test $H_{03}$: $\beta_3 = 0$ in the cubic model

$$y = \beta_0 + \beta_1 X + \beta_2 X^2 + \beta_3 X^3 + \epsilon$$

If one also rejects $H_{03}$: $\beta_3 = 0$, then use the cubic model or consider using a nonlinear model, otherwise use the quadratic model. If one fails to reject $H_{02}$: $\beta_2 = 0$ in the quadratic model, then one should test $H_{01}$: $\beta_1 = 0$ in the simple linear model

$$y = \beta_0 + \beta_1 X + \epsilon$$

If $H_{01}$: $\beta_1 = 0$ is rejected, then the model selected should be the simple linear model. If $H_0$: $\beta_1 = 0$ cannot be rejected, then one is not able to say the different treatments have any effect.

If the trend appears to be curvilinear, then one should begin by fitting a cubic model

$$y = \beta_0 + \beta_1 X + \beta_2 X^2 + \beta_3 X^3 + \epsilon$$

to the data and test $H_{03}$: $\beta_3 = 0$. If $H_{03}$: $\beta_3 = 0$ is rejected, conclude the appropriate model is cubic, provided that a plot of the model appears to model the data adequately, otherwise consider using a nonlinear model. If one fails to reject $H_{03}$: $\beta_3 = 0$, then one should test $H_{02}$: $\beta_2 = 0$ in the quadratic model. If $H_{02}$: $\beta_2 = 0$ is rejected, conclude the model is quadratic. If one also fails to reject $H_{02}$: $\beta_2 = 0$, then one should test $H_{01}$: $\beta_1 = 0$ in the simple linear model and finish as in the preceding paragraph.

The above method of attack should provide some information about the effects of quantitative factors. If none of these models seems to be adequate, one should probably consider a nonlinear model to model the data rather than increasing the degree of the polynomial model. See Gallant (1987) for a discussion of nonlinear models. For nonlinear models to be an improvement over polynomial models, one will probably need more than eight or nine levels of the treatment factor.

This chapter concludes with an example.

**10.3
AN EXAMPLE**

The observed means for firmness for the potato example described earlier are given as follows:

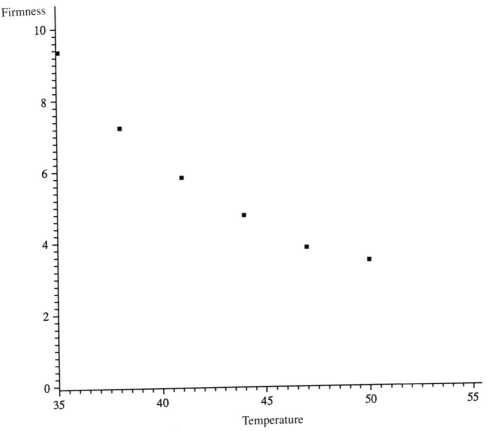

**Figure 10.2   A Plot of the Data in Example 10.3.**

| Temperature | 35   | 38  | 41  | 44  | 47  | 50  |
|-------------|------|-----|-----|-----|-----|-----|
| Firmness    | 15.3 | 8.6 | 4.7 | 3.6 | 3.5 | 3.6 |

A plot of the data is given in Figure 10.2.

Examination of the plot in Figure 10.2 reveals a curvilinear trend in the data. So we begin by fitting the cubic model

$$\text{Firmness} = \beta_0 + \beta_1(\text{Temp}) + \beta_2(\text{Temp})^2 + \beta_3(\text{Temp})^3 + \epsilon \quad (10.3.1)$$

The analysis of the model in (10.3.1) gave the estimates and their associated $t$-statistics and $p$-values in Table 10.1.

The coefficient of the cubic term in Table 10.1 is not significantly different from zero ($P = .4133$). Hence, one should fit the quadratic model,

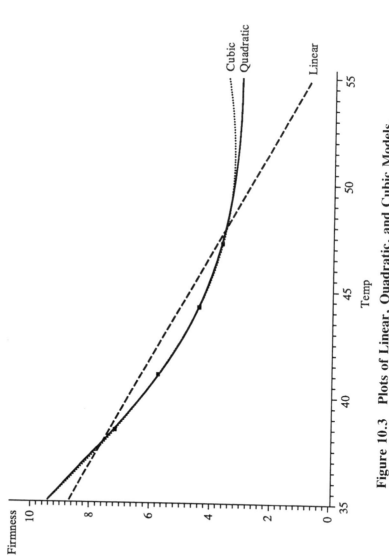

**Figure 10.3** Plots of Linear, Quadratic, and Cubic Models.

**Table 10.1  Parameter Estimates for a Cubic Model**

| Parameter | Estimate | T for $H_0$: Parameter $=0$ | $PR > \mid T \mid$ | STD Error of Estimate |
|---|---|---|---|---|
| Intercept | 84.88183226 | 3.48 | 0.0736 | 24.40139381 |
| Temp | −3.97710905 | −2.28 | 0.1507 | 1.74772994 |
| Temp * Temp | 0.06358319 | 1.54 | 0.2644 | 0.04140974 |
| Temp * Temp * Temp | −0.00033265 | −1.02 | 0.4133 | 0.00032463 |

**Table 10.2  Parameter Estimates for a Quadratic Model**

| Parameter | Estimate | T for $H_0$: Parameter $=0$ | $PR > \mid T \mid$ | STD Error of Estimate |
|---|---|---|---|---|
| Intercept | 59.98849206 | 25.91 | 0.0001 | 2.31554649 |
| Temp | −2.18969444 | −19.89 | 0.0003 | 0.11011099 |
| Temp * Temp | 0.02117063 | 16.36 | 0.0005 | 0.00129371 |

$$\text{Firmness} = \beta_0 + \beta_1(\text{Temp}) + \beta_2(\text{Temp})^2 + \epsilon \qquad (10.3.2)$$

to the data. The analysis of the model in (10.3.2) gave the estimates and their associated $t$-statistics and $p$-values in Table 10.2.

The coefficient for the quadratic term in Table 10.2 is significantly different from zero ($P = .0005$). Hence the quadratic model is chosen to model the data. The fitted model is

$$\text{Firmness} = 59.988 - 2.190(\text{Temp}) + .022(\text{Temp})^2$$

The data and the quadratic model are plotted in Figure 10.3. For comparison purposes the best cubic model and the best linear model are also plotted. These plots also make apparent the dangers of extrapolating beyond the range of the data when using polynomial models, particularly when the wrong form of the model is selected.

In the next chapter the methods discussed in this chapter are generalized to include the case where there are several quantitative factors which are being varied.

# 11

# QUADRATIC RESPONSE SURFACE MODELS

**CHAPTER OUTLINE**

C onsider a two-way treatment structure where both treatment factors are quantitative. Suppose the levels of factor 1 are given by

$$X_{11}, X_{12}, \ldots, X_{1t}$$

and that the levels of factor 2 are given by

$$X_{21}, X_{22}, \ldots, X_{2b}$$

The polynomial models described in Chapter 10 can be generalized to help analyze nonreplicated two-way experiments and higher-order cross-classified treatment structures when all treatment factors are quantitative. These models are especially useful when one of the important goals of the experiment is to determine combinations of the two treatment factors which will give near-optimum responses. The polynomial models described in this chapter have been so successful in this regard that many statisticians do not even recommend complete replications in these kinds of experiments. In fact, if one chooses the design points (i.e., the combinations of the two factors with which to run experiments) carefully, polynomial models do not even require one to have a complete replicate. Many special treatment designs exist which will help researchers gain valuable information with modest numbers of experimental runs. See for example Box, Hunter, and Hunter (1978) and Cochran and Cox (1957).

For the two-way experiment described above, a quadratic response surface model will generally provide an adequate fit to the data. This model is defined by

$$y_{ij} = \beta_0 + \beta_1 X_{1i} + \beta_2 X_{2j} + \beta_3 X_{1i}^2 + \beta_4 X_{1i}X_{2j} + \beta_5 X_{2j}^2 + \epsilon_{ij}$$

$$i = 1,2, \ldots, t, \quad j = 1,2, \ldots, b$$

This model is generalized in the next section to include the case when there are $k$ quantitative factors affecting the response variables.

## 11.1 ANALYZING A QUADRATIC RESPONSE SURFACE MODEL

Consider an experiment where one desires to study the effects of $k$ different factors, $X_1, X_2, \ldots, X_k$ on a dependent variable $y$. Suppose data are collected from experimental runs at the $n$ design points $(X_{11}, X_{21}, \ldots, X_{k1}), (X_{12}, X_{22}, \ldots, X_{k2}), \ldots, (X_{1n}, X_{2n}, \ldots, X_{kn})$. The quadratic response surface model for this case is

$$
\begin{aligned}
y_i = \beta_{00} &+ \beta_{10}X_{1i} + \beta_{20}X_{2i} + \cdots + \beta_{k0}X_{ki} \\
&+ \beta_{11}X_{1i}^2 + \beta_{22}X_{2i}^2 + \cdots + \beta_{kk}X_{ki}^2 \\
&+ \beta_{12}X_{1i}X_{2i} + \beta_{13}X_{1i}X_{3i} + \cdots + \beta_{1k}X_{1i}X_{ki} \\
&+ \beta_{23}X_{2i}X_{3i} + \cdots + \beta_{k-1k}X_{k-1i}X_{ki} + \epsilon_i
\end{aligned}
$$

$$i = 1, 2, \ldots, n \qquad (11.1.1)$$

Any statistical computing software which will fit multiple regression models can be used to fit model (11.1.1). Some software packages such as SAS® have special programs for response surface models. Although such programs are not necessary, they do provide output in a form that is quite useful for interpreting the observed results of the experiment. This is illustrated by an example in the next section.

**11.2
AN EXAMPLE**

An experiment was conducted to study the effects of three factors on the quality of frankfurters. The three factors to be studied were (1) sodium acid pyrophosphate (SAPP), (2) sodium tripolyphosphate (STPP), and (3) sodium chloride (NACL). Measures of quality taken on the finished product included (a) saltiness (SALT), (b) soapiness (SOAP), and (c) texture (TEX). The treatment combinations (i.e., design points) which were used to make frankfurters were those given by a "central rotatable response surface design with center points." Such designs are tabled in Cochran and Cox (1957) and usually provide an optimal set of design points which do not require a large number of runs. The treatment combinations given by this design are listed in Table 11.1 under the

**Table 11.1   Treatment Combinations and Data
for Response Surface Example**

| TRT | NACL | SAPP | STPP | SALT | SOAP | TEX |
|---|---|---|---|---|---|---|
| 1 | 1.60 | 0.050 | 0.10 | 3.20 | 0.65 | 2.00 |
| 2 | 1.90 | 0.050 | 0.10 | 7.00 | 9.00 | 4.85 |
| 3 | 1.60 | 0.050 | 0.40 | 1.20 | 7.90 | 2.30 |
| 4 | 1.90 | 0.050 | 0.40 | 4.45 | 4.85 | 3.40 |
| 5 | 1.60 | 0.200 | 0.10 | 4.40 | 0.80 | 5.35 |
| 6 | 1.90 | 0.200 | 0.10 | 2.85 | 3.15 | 4.30 |
| 7 | 1.60 | 0.200 | 0.40 | 0.15 | 5.80 | 4.10 |
| 8 | 1.90 | 0.200 | 0.40 | 5.55 | 4.00 | 4.20 |
| 9 | 1.50 | 0.125 | 0.25 | 1.00 | 4.75 | 5.85 |
| 10 | 2.00 | 0.125 | 0.25 | 5.75 | 3.35 | 6.85 |
| 11 | 1.75 | 0.125 | 0.00 | 4.90 | 1.20 | 2.35 |
| 12 | 1.75 | 0.125 | 0.50 | 3.60 | 4.65 | 4.35 |
| 13 | 1.75 | 0.000 | 0.25 | 3.60 | 5.30 | 1.10 |
| 14 | 1.75 | 0.250 | 0.25 | 5.00 | 4.60 | 3.40 |
| 15 | 1.75 | 0.125 | 0.25 | 4.00 | 9.00 | 3.20 |
| 16 | 1.75 | 0.125 | 0.25 | 4.20 | 8.00 | 3.55 |
| 17 | 1.75 | 0.125 | 0.25 | 1.85 | 7.30 | 4.20 |
| 18 | 1.75 | 0.125 | 0.25 | 3.45 | 7.65 | 4.25 |
| 19 | 1.75 | 0.125 | 0.25 | 5.30 | 6.90 | 4.20 |
| 20 | 1.75 | 0.125 | 0.25 | 5.35 | 7.80 | 4.50 |

columns labeled NACL, SAPP, and STPP. A completely new product was made (in a random order) with each of the (NACL, SAPP, STPP) combinations listed in Table 11.1. The resulting products were given independent evaluations by each of 10 judges. Since the judges' scores do not provide independent replications of the frankfurters, the mean of the judges' scores on each quality variable is the appropriate variable for analysis. These means are also given in Table 11.1 under the columns labeled SALT, SOAP, and TEX.

Each of the response variables in Table 11.1 was modeled by a quadratic response surface model of the form:

$$y = \beta_0 + \beta_1(NACL) + \beta_2(SAPP) + \beta_3(STPP)$$
$$+ \beta_4(NACL)^2 + \beta_5(SAPP)^2 + \beta_6(STPP)^2$$
$$+ \beta_7(NACL)(SAPP) + \beta_8(NACL)(STPP)$$
$$+ \beta_9(SAPP)(STPP) + \epsilon$$

where $y$ is any one of the three response variables, SALT, SOAP, or TEX.

Each of the response variables was analyzed by the above model and the resulting estimates of the parameters are given in Table 11.2. The squared multiple correlation coefficient ($R^2$) for each model is also given.

Quadratic response surface models are not always easy to interpret. In order to interpret the results, it is often helpful to

## Table 11.2 Response Surface Parameter Estimates

| | | ESTIMATES | | |
|---|---|---|---|---|
| *Model Term* | *Parameter* | *SALT* | *SOAP* | *TEX* |
| Intercept | $\beta_0$ | −45.01 | −208.39 | 93.64 |
| NACL | $\beta_1$ | 53.74 | 209.88 | −116.83 |
| SAPP | $\beta_2$ | 52.51 | 112.08 | 134.30 |
| STPP | $\beta_3$ | −70.46 | 189.30 | 12.41 |
| (NACL)$^2$ | $\beta_4$ | −13.98 | −51.42 | 36.22 |
| (SAPP)$^2$ | $\beta_5$ | 3.28 | −148.07 | −117.53 |
| (STPP)$^2$ | $\beta_6$ | .02 | −69.42 | −11.78 |
| NACL * SAPP | $\beta_7$ | −35.56 | −52.78 | −54.44 |
| NACL * STPP | $\beta_8$ | 35.56 | −86.39 | −3.33 |
| SAPP * STPP | $\beta_9$ | 33.33 | 30.56 | −2.22 |
| $R^2$ | | .69 | .86 | .85 |

1. Determine the effect of each of the treatment factors after adjusting for the effects of all of the other treatment factors.

2. Determine the portion of the overall $R^2$ that is due to (a) the linear terms in the model, (b) the quadratic terms in the model after adjusting for the linear terms, and (c) the cross-product terms in the model after adjusting for both linear and quadratic terms.

3. Construct contour plots for each of the fitted models.

Many statistical computing packages do combinations of these three things automatically. SAS® RSREG, for example, does (1) and (2) automatically, and X-STAT, for example, does (3) very easily.

Table 11.3 shows what portion of the $R^2$ values in Table 11.2 are due to linear effects, quadratic effects adjusted for linear effects, and cross-product effects after adjusting for both linear and quadratic effects. Table 11.3 also includes $p$-values for the hypotheses that these partitions of the overall model sum of squares are significantly different from zero.

Table 11.4 shows the results of $F$-tests for determining the effects of each of the treatment factors after adjusting for all of the effects of the other two factors for the data in Table 11.1.

Examination of the results in Tables 11.2, 11.3, and 11.4 reveals that

1. The only factor significantly affecting saltiness is NACL. The quadratic model obtained by regressing SALT on NACL only is given by

$$SALT = -55.17 + 58.43(NACL) - 14.05(NACL)^2$$

in which the quadratic term was not significant. The simple linear model obtained by regressing SALT on NACL is

$$SALT = -12.35 + 9.25(NACL)$$

**Table 11.3  Partitions of $R^2$ Values**

| Effects | SALT $R^2$ | SALT p-value | SOAP $R^2$ | SOAP p-value | TEX $R^2$ | TEX p-value |
|---|---|---|---|---|---|---|
| Linear | .533 | .016 | .184 | .033 | .233 | .021 |
| Quadratic | .024 | .854 | .415 | .003 | .528 | .001 |
| Cross product | .129 | .310 | .261 | .012 | .089 | .180 |
| Total $R^2$ | .686 | .092 | .860 | .003 | .850 | .004 |

**Table 11.4  *F*-Tests for the Treatment Factors**

| | | SALT | | SOAP | | TEX | |
| --- | --- | --- | --- | --- | --- | --- | --- |
| *Factor* | *df* | *F* | *p-value* | *F* | *p-value* | *F* | *p-value* |
| NACL | 4 | 4.60 | .023 | 7.23 | .005 | 6.81 | .007 |
| SAPP | 4 | 0.33 | .851 | 2.83 | .083 | 7.53 | .005 |
| STPP | 4 | 1.53 | .266 | 11.15 | .001 | .53 | .72 |

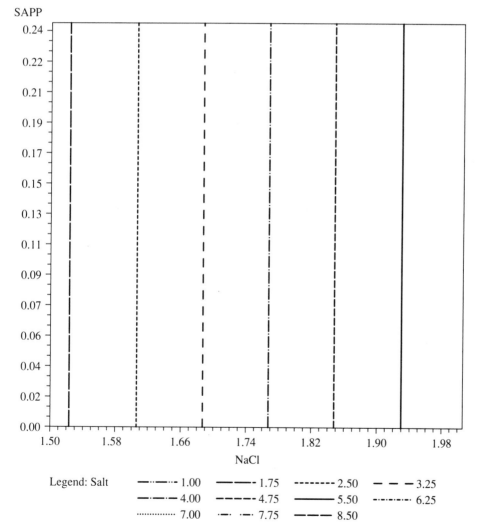

**Figure 11.1   Contour Plot for Saltiness.**

2. All three treatment factors affect soapiness, and since both quadratic terms and cross-product terms are significant in Table 11.3, the model for SOAP cannot be simplified.

3. Only NACL and SAPP affect texture. The quadratic response surface model obtained by regressing TEX on NACL and SAPP only is given by

$$TEX = 99.15 - 121.41(NACL) + 132.67(SAPP)$$
$$+ 37.29(NACL)^2 - 54.44(NACL)(SAPP)$$
$$- 113.24(SAPP)^2$$

for which all coefficients are significantly different from zero.

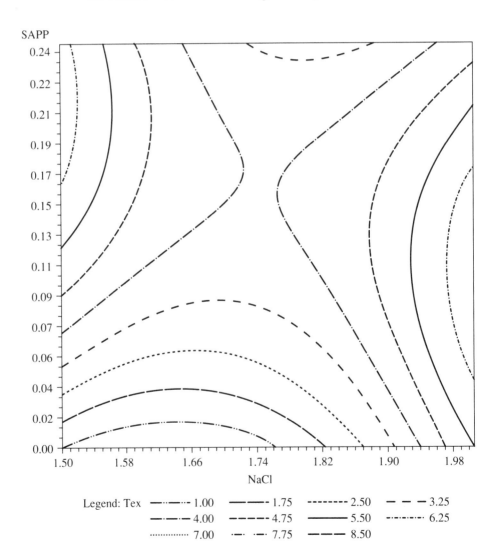

**Figure 11.2   Contour Plot for Texture.**

Before constructing contour plots, we first note that the effect of NACL on saltiness is essentially the same for all levels of the other two factors, SAPP and STPP, and the effect of NACL and SAPP on texture is essentially the same for all levels of STPP. Since STPP does not affect either saltiness or texture, we will construct contour plots for saltiness and texture with STPP = 0, and since STPP does affect soapiness, we will construct contours for soapiness at several different values of STPP.

The contours for saltiness and texture are given in Figures 11.1 and 11.2, respectively. Note that a particular contour shows all of the values of the treatment factors corresponding to each axis for which the predicted response is constant. Each contour represents a different constant.

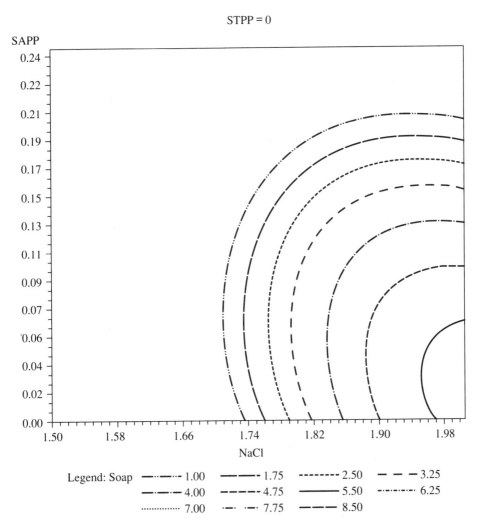

**Figure 11.3   Contour Plot for Soapiness When STPP = 0.**

The contour plot in Figure 11.1 is simply a series of vertical lines. This is because saltiness was a simple linear function of NACL only and did not depend on SAPP. The dashed line at NACL = 1.93 corresponds to a contour value of 5.50. The interpretation of the contour is that NACL = 1.93 gives a predicted value of 5.50 for saltiness for all possible values of SAPP. Recall that the linear equation relating SALT to NACL was

$$SALT = -12.35 + 9.25(NACL)$$

and evaluating this equation when NACL = 1.93, we get

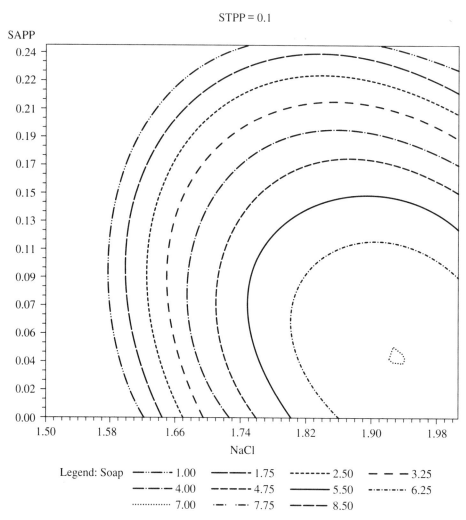

Figure 11.4    Contour Plot for Soapiness When STPP = .1.

$$\text{SALT} = -12.35 + (9.25)(1.93)$$
$$= -12.35 + 17.85 = 5.50$$

Examination of the contour plot in Figure 11.1 also reveals that if one wanted to obtain a product which had a value of SALT which is no more than 3.25, then the amount of NACL which could be used would need to be less than 1.69.

The contour plot for texture in Figure 11.2 is much more complex and certainly more interesting. The solid line corresponds to a contour value of 5.50. Any combination of NACL and SAPP which falls on this

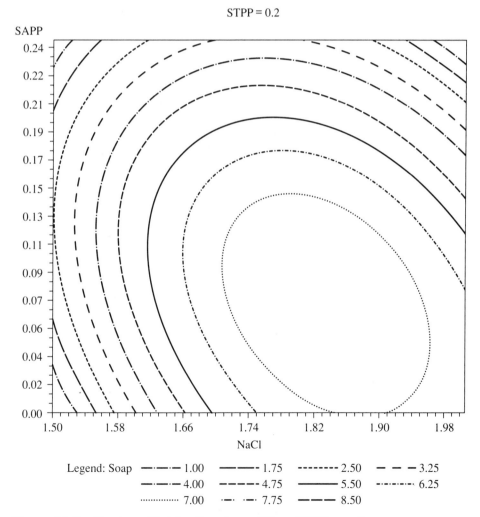

STPP = 0.2

Legend: Soap
—·—·— 1.00    ———· 1.75    -------- 2.50    — — — 3.25
—·—·— 4.00    - - - - - 4.75    ——— 5.50    ·—··—··— 6.25
············ 7.00    —· ··— 7.75    ——— 8.50

**Figure 11.5   Contour Plot for Soapiness When STPP = .2.**

contour should provide a texture value close to 5.50. For example, SAPP = .17 and NACL = 1.55 or NACL = 1.95 should both give a texture value close to 5.50.

Soapiness depends on all three treatment factors, thus a single contour plot will not usually provide enough information to accurately assess the effect of all three factors. Because of this, contour plots were constructed for several different values of STPP, namely, STPP = 0, .1, .2, .3, .4, and .5. These contour plots are shown in Figures 11.3 to 11.8, respectively.

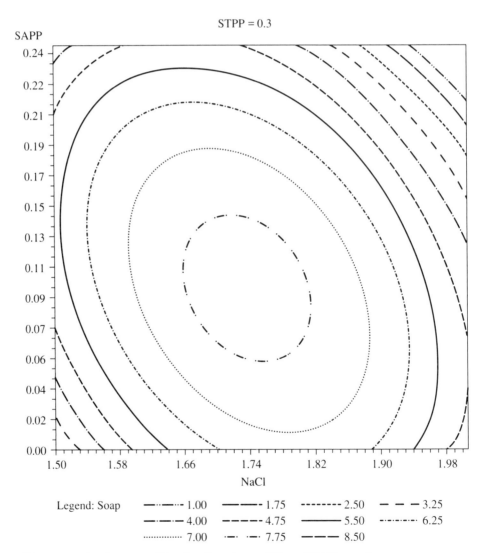

Figure 11.6   Contour Plot for Soapiness When STPP = .3.

The solid line in each plot shows those treatment factor combinations where SOAP = 5.50. For example,

$$STPP = 0, \quad SAPP = .04,$$

$$and \quad NACL = 1.95$$

gives a predicted response of 5.50 for soapiness as does

$$STPP = .4, \quad SAPP = .15,$$

$$and \quad NACL = 1.84$$

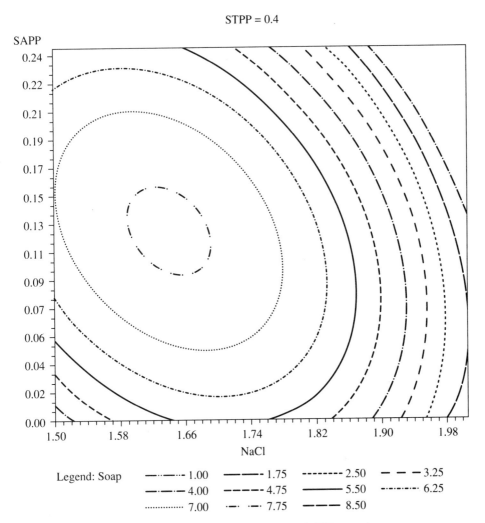

**Figure 11.7   Contour Plot for Soapiness When STPP = .4.**

Suppose now that the objective is to find combinations of the treatment factors which would be expected to produce very good products. In most instances it is unlikely that any given combination of the treatment factors will provide an optimum response for all dependent variables, and one usually has to compromise somewhat. Suppose, for the example, it is desired to have a texture value as large as possible, but not over 5.5, a soapiness value as small as possible, and a saltiness value as small as possible. Further suppose that the product is acceptable to most consumers if SOAP ≤ 2.5 and SALT ≤ 3.25. What combinations of the treatment factors should produce acceptable products? To answer this question, it is helpful to make transparencies of each of the plots

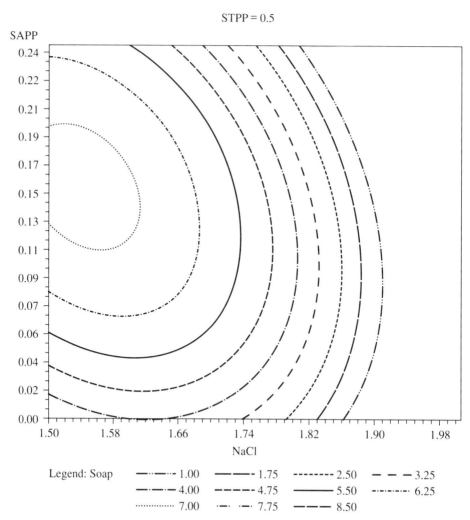

**Figure 11.8   Contour Plot for Soapiness When STPP = .5.**

in Figures 11.1 to 11.8 so that these plots can be overlayed to see the effects of different treatment combinations simultaneously.

Overlaying Figure 11.1 on Figure 11.2 reveals the combinations of NACL and SAPP which will produce frankfurters which have SALT ≤ .25 and 4.00 ≤ TEX ≤ 5.5. This region is shown in Figure 11.9 and includes the regions labeled *A*, *B*, and *C*. Overlaying Figure 11.9 on Figures 11.3 to 11.8 will reveal which of these combinations will also have SOAP ≤ 2.5.

Overlaying Figure 11.9 on Figure 11.3 shows that all of the acceptable combinations of NACL and SAPP in the regions labeled *A*, *B*, and *C* of Figure 11.9 when combined with STPP = 0 should also produce a product which is acceptable on soapiness. Overlaying Figure 11.9 on Figure 11.4 shows that a few of those combinations acceptable when STPP = 0 are no longer acceptable when STPP = .1;

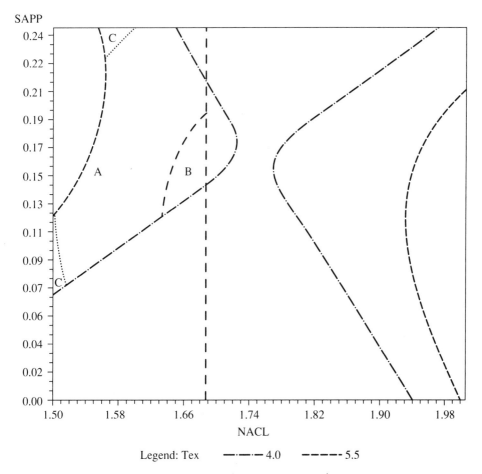

**Figure 11.9   Acceptable Region for Saltiness and Texture.**

the combinations which are now unacceptable are those in the region labeled B. Overlaying Figure 11.9 on Figure 11.5 shows that when STPP = .2, the only combinations of the other two factors which should still produce acceptable products are those in the regions labeled by $C$. Overlaying Figure 11.9 on Figures 11.6 to 11.8 shows that when STPP $\geq$ .3, there are no combinations of NACL and SAPP which would be expected to produce a frankfurter which has SOAP $\leq$ 2.5.

After using response surface methods to determine acceptable combinations of the treatments factors, one should always test these combinations by making new products. This will enable one to: (1) verify that these combinations do, in fact, produce acceptable products, and (2) determine which of the acceptable combinations produces the best product.

# Appendices

## Table A.1 Critical Points for $\ell_1/(\ell_1 + \ell_2 + \cdots + \ell_p)$

| n | \| | | | | | p | | | | |
|---|---|---|---|---|---|---|---|---|---|---|
| | 2 | 3 | 4 | 5 | 6 | 7 | 9 | 11 | 15 | 19 |

UPPER 10%

| n | 2 | 3 | 4 | 5 | 6 | 7 | 9 | 11 | 15 | 19 |
|---|---|---|---|---|---|---|---|---|---|---|
| 2 | .998* | | | | | | | | | |
| 3 | .974* | .834 | | | | | | | | |
| 4 | .943* | .846* | .802 | | | | | | | |
| 5 | .914* | .813 | .698 | .640 | | | | | | |
| 6 | .888* | .763* | .655 | .636 | .569 | | | | | |
| 7 | .866* | .744 | .649 | .573 | .546 | .510 | | | | |
| 9 | .831* | .675 | .606 | .557 | .497 | .449 | .398 | | | |
| 11 | .804* | .659 | .570 | .500 | .456 | .429 | .383 | .351 | | |
| 15 | .765* | .602 | .513 | .455 | .414 | .382 | .327 | .295 | .256 | |
| 19 | .738* | .574 | .484 | .433 | .378 | .350 | .302 | .277 | .233 | .205 |
| 31 | .689* | .516 | .431 | .374 | .333 | .299 | .256 | .226 | .188 | .165 |
| 49 | .651* | .489 | .398 | .336 | .291 | .262 | .223 | .196 | .160 | .139 |
| 99 | .607* | .442 | .349 | .294 | .253 | .226 | .185 | .160 | .128 | .100 |

UPPER 5%

| n | 2 | 3 | 4 | 5 | 6 | 7 | 9 | 11 | 15 | 19 |
|---|---|---|---|---|---|---|---|---|---|---|
| 2 | .999* | | | | | | | | | |
| 3 | .987* | .857 | | | | | | | | |
| 4 | .965* | .882* | .841 | | | | | | | |
| 5 | .941* | .851 | .759* | .668 | | | | | | |
| 6 | .917* | .801* | .682 | .657* | .596 | | | | | |
| 7 | .897* | .781 | .683* | .599 | .575* | .535 | | | | |
| 9 | .863* | .704 | .632* | .590 | .522* | .469 | .414 | | | |
| 11 | .836* | .694 | .595* | .524 | .484* | .450 | .402 | .367 | | |

## Table A.1   (Continued)

| | | | | | $p$ | | | | | |
|---|---|---|---|---|---|---|---|---|---|---|
| $n$ | 2 | 3 | 4 | 5 | 6 | 7 | 9 | 11 | 15 | 19 |

UPPER 5%

| | | | | | | | | | | |
|---|---|---|---|---|---|---|---|---|---|---|
| 15 | .795* | .630 | .543* | .476 | .434* | .399 | .339 | .306 | .266 | |
| 19 | .766* | .629 | .509* | .454 | .400* | .365 | .314 | .288 | .241 | .212 |
| 31 | .713* | .535 | .450* | .389 | .344* | .311 | .264 | .234 | .194 | .170 |
| 49 | .671* | .508 | .407* | .348 | .300 | .271 | .230 | .202 | .164 | .142 |
| 99 | .622* | .461 | .358 | .302 | .260 | .231 | .189 | .163 | .131 | .112 |

UPPER 1%

| | | | | | | | | | | |
|---|---|---|---|---|---|---|---|---|---|---|
| 2 | .99997* | | | | | | | | | |
| 3 | .998* | .893 | | | | | | | | |
| 4 | .988* | .934* | .900 | | | | | | | |
| 5 | .974* | .908 | .825* | .719 | | | | | | |
| 6 | .959* | .863 | .733 | .721* | .646 | | | | | |
| 7 | .943* | .845 | .748* | .647 | .624 | .581 | | | | |
| 9 | .914* | .758 | .693* | .652 | .575* | .504 | .445 | | | |
| 11 | .888* | .741 | .651* | .570 | .532* | .491 | .437 | .397 | | |
| 15 | .847* | .626 | .593* | .517 | .474* | .433 | .363 | .337 | .284 | |
| 19 | .816* | .597 | .554* | .496 | .436* | .395 | .336 | .310 | .257 | .225 |
| 31 | .757* | .537 | .485* | .420 | .371* | .333 | .282 | .249 | .205 | .180 |
| 49 | .709* | .504 | .435* | .372 | .318 | .286 | .243 | .214 | .173 | .150 |
| 99 | .650* | .446 | .377 | .319 | .272 | .242 | .198 | .170 | .136 | .116 |

*Indicates an exact critical point.

## Table A.2 Critical Points for $\ell_2/(\ell_2 + \ell_3 + \cdots + \ell_p)$

| | | | | $p$ | | | |
|---|---|---|---|---|---|---|---|
| | | | | UPPER 10% | | | |
| $n$ | 3 | 4 | 5 | 6 | 7 | 8 | 9 |
| 3 | .997 | | | | | | |
| 4 | .971 | .899 | | | | | |
| 5 | .937 | .825 | .750 | | | | |
| 6 | .901 | .778 | .693 | .634 | | | |
| 7 | .874 | .734 | .643 | .583 | .550 | | |
| 8 | .849 | .702 | .621 | .552 | .500 | .487 | |
| 9 | .833 | .685 | .592 | .531 | .492 | .452 | .428 |
| | | | | UPPER 5% | | | |
| 3 | .9993 | | | | | | |
| 4 | .986 | .930 | | | | | |
| 5 | .970 | .860 | .797 | | | | |
| 6 | .927 | .806 | .728 | .675 | | | |
| 7 | .909 | .782 | .682 | .610 | .575 | | |
| 8 | .890 | .741 | .656 | .575 | .542 | .509 | |
| 9 | .860 | .714 | .646 | .562 | .511 | .476 | .442 |
| | | | | UPPER 1% | | | |
| 3 | .99994 | | | | | | |
| 4 | .997 | .968 | | | | | |
| 5 | .995 | .922 | .851 | | | | |
| 6 | .967 | .863 | .826 | .739 | | | |
| 7 | .965 | .854 | .752 | .667 | .633 | | |
| 8 | .942 | .790 | .721 | .639 | .592 | .561 | |
| 9 | .913 | .775 | .687 | .626 | .557 | .522 | .491 |

## Table A.3 Expected Values of $\ell_i / \sigma^2$ for $i = 1, 2, 3$

| | | | | | $p$ | | | | |
|---|---|---|---|---|---|---|---|---|---|
| $i$ | $n$ | 2 | 3 | 4 | 5 | 6 | 7 | 8 | 9 |
| 1 | 2 | 3.57* | | | | | | | |
| | 3 | 5.00* | 6.85 | | | | | | |
| | 4 | 6.36* | 8.33 | 10.21 | | | | | |
| | 5 | 7.33* | 9.99 | 11.71 | 13.57 | | | | |
| | 6 | 8.94* | 11.22 | 13.54 | 15.49 | 17.22 | | | |
| | 7 | 10.20* | 12.72 | 14.96 | 16.91 | 19.19 | 20.81 | | |
| | 8 | 11.42 | 13.98 | 16.54 | 18.80 | 20.59 | 23.14 | 25.02 | |
| | 9 | 12.66* | 15.51 | 18.29 | 20.10 | 22.59 | 24.71 | 26.42 | 28.21 |
| 2 | 2 | 0.43* | | | | | | | |
| | 3 | 1.00* | 2.03 | | | | | | |
| | 4 | 1.64* | 3.00 | 4.32 | | | | | |
| | 5 | 2.67* | 4.02 | 5.45 | 6.87 | | | | |
| | 6 | 3.06* | 4.98 | 6.81 | 8.28 | 9.74 | | | |
| | 7 | 3.80* | 5.96 | 7.94 | 9.50 | 11.27 | 12.76 | | |
| | 8 | 4.55 | 6.90 | 9.18 | 10.96 | 12.61 | 14.47 | 16.20 | |
| | 9 | 5.34* | 8.04 | 10.31 | 12.23 | 14.10 | 15.96 | 17.47 | 19.21 |
| 3 | 3 | | 0.26 | | | | | | |
| | 4 | | 0.66 | 1.41 | | | | | |
| | 5 | | 1.14 | 2.19 | 3.19 | | | | |
| | 6 | | 1.67 | 3.08 | 4.23 | 5.50 | | | |
| | 7 | | 2.27 | 3.84 | 5.21 | 6.61 | 7.74 | | |
| | 8 | | 2.79 | 4.76 | 6.20 | 7.74 | 9.06 | 10.54 | |
| | 9 | | 3.55 | 5.50 | 7.28 | 8.86 | 10.35 | 11.73 | 13.18 |

*These values are exact.

### Table A.4  Upper Critical Points for $F_T$ and $F_B$

| | | | | $p$ | | | | |
|---|---|---|---|---|---|---|---|---|
| $n$ | 2 | 3 | 4 | 5 | 6 | 7 | 8 | 9 |

UPPER 10%

| | | | | | | | | |
|---|---|---|---|---|---|---|---|---|
| 2 | 63.0 | 9.00 | 9.99 | 9.82 | 3.84 | 3.63 | 3.21 | 3.02 |
| 3 | 9.17 | 3.91 | 3.09 | 3.04 | 2.70 | 2.63 | 2.52 | 2.47 |
| 4 | 5.54 | 3.10 | 2.80 | 2.31 | 2.51 | 2.25 | 2.21 | 2.26 |
| 5 | 4.43 | 2.65 | 2.29 | 2.33 | 2.12 | 2.04 | 2.05 | 1.95 |
| 6 | 3.39 | 2.37 | 2.08 | 2.03 | 2.01 | 1.94 | 1.96 | 1.82 |
| 7 | 2.94 | 2.37 | 2.03 | 1.93 | 1.82 | 1.96 | 1.80 | 1.83 |
| 8 | 2.83 | 2.19 | 1.93 | 1.89 | 1.89 | 1.79 | 1.83 | 1.84 |
| 9 | 2.34 | 2.05 | 1.93 | 1.77 | 1.83 | 1.81 | 1.71 | 1.75 |

UPPER 5%

| | | | | | | | | |
|---|---|---|---|---|---|---|---|---|
| 2 | 219.1 | 19.0 | 11.91 | 7.69 | 5.30 | 4.96 | 4.44 | 4.30 |
| 3 | 19.17 | 6.38 | 4.59 | 4.33 | 4.01 | 3.65 | 3.20 | 3.17 |
| 4 | 13.74 | 4.14 | 3.72 | 2.92 | 3.26 | 2.87 | 2.84 | 2.94 |
| 5 | 6.31 | 3.46 | 2.78 | 2.80 | 2.57 | 2.62 | 2.49 | 2.38 |
| 6 | 4.80 | 3.24 | 2.66 | 2.49 | 2.32 | 2.52 | 2.31 | 2.22 |
| 7 | 4.23 | 2.94 | 3.07 | 2.29 | 2.25 | 2.34 | 2.15 | 2.18 |
| 8 | 3.66 | 2.68 | 2.37 | 2.27 | 2.20 | 2.11 | 2.16 | 2.09 |
| 9 | 3.07 | 2.56 | 2.32 | 2.08 | 2.13 | 2.10 | 2.05 | 2.02 |

UPPER 1%

| | | | | | | | | |
|---|---|---|---|---|---|---|---|---|
| 2 | 5030.00 | 99.00 | 53.86 | 19.77 | 12.98 | 10.39 | 9.08 | 8.22 |
| 3 | 98.83 | 15.20 | 9.53 | 6.62 | 6.69 | 5.73 | 4.78 | 4.94 |
| 4 | 27.47 | 6.72 | 6.39 | 4.38 | 6.04 | 3.85 | 3.63 | 4.35 |
| 5 | 16.71 | 6.05 | 3.77 | 4.61 | 3.87 | 4.00 | 3.47 | 3.41 |
| 6 | 9.08 | 5.46 | 4.37 | 3.69 | 3.82 | 3.86 | 3.20 | 3.16 |
| 7 | 8.79 | 4.94 | 3.86 | 3.41 | 3.26 | 3.63 | 3.10 | 3.04 |
| 8 | 7.02 | 4.22 | 3.45 | 3.02 | 2.94 | 2.79 | 2.75 | 2.83 |
| 9 | 5.40 | 3.40 | 3.20 | 2.91 | 3.10 | 2.90 | 2.69 | 2.65 |

*Note:* The critical points above the diagonal are used when the degrees of freedom for the numerator mean square is equal to $p$. Those below the diagonal are used when the degrees of freedom of the numerator mean square is equal to $n$.

### Table A.5  The 1% Critical Points for $\Lambda$ When $q = 1$ (the critical points for $\delta = 9$ are approximately correct for values of $\delta > 9$)

| | | | | $b$ | | | |
|---|---|---|---|---|---|---|---|
| $t$ | $\delta$ | 3 | 4 | 5 | 6 | 8 | 10 |
| 3 | 0 | $.0^5436$ | $.0^3931$ | .00687 | .0185 | .0422 | .0812 |
| | 1 | $.0^5768$ | .00153 | .00917 | .0233 | .0530 | .0914 |
| | 2 | $.0^4290$ | .00488 | .0196 | .0446 | .0937 | .1304 |
| | 3 | $.0^4984$ | .0130 | .0426 | .0897 | .1743 | .2098 |
| | 4 | $.0^3188$ | .0198 | .0659 | .1328 | .2640 | .3103 |
| | 5 | $.0^3231$ | .0212 | .0750 | .1482 | .3076 | .3800 |
| | 6 | $.0^3240$ | .0208 | .0755 | .1488 | .3059 | .4027 |
| | 7 | $.0^3240$ | .0206 | .0748 | .1475 | .2924 | .4031 |
| | 8 | $.0^3239$ | .0205 | .0744 | .1470 | .2840 | .4004 |
| | 9 | $.0^3239$ | .0205 | .0744 | .1469 | .2807 | .3988 |
| 4 | 0 | .00304 | .0374 | .0902 | .1455 | .2512 | .3322 |
| | 1 | .00388 | .0470 | .1025 | .1578 | .2645 | .3449 |
| | 2 | .00702 | .0789 | .1436 | .1974 | .3066 | .3852 |
| | 3 | .0124 | .1232 | .2093 | .2625 | .3764 | .4520 |
| | 4 | .0166 | .1498 | .2666 | .3324 | .4553 | .5314 |
| | 5 | .0181 | .1545 | .2920 | .3781 | .5139 | .5972 |
| | 6 | .0183 | .1529 | .2977 | .2945 | .5406 | .6319 |
| | 7 | .0184 | .1520 | .2979 | .3963 | .5471 | .6417 |
| | 8 | .0183 | .1516 | .2977 | .3951 | .5471 | .6414 |
| | 9 | .0183 | .1516 | .2977 | .3943 | .5463 | .6396 |
| 5 | 0 | .0252 | .1232 | .2241 | .3148 | .4215 | .4930 |
| | 1 | .0260 | .1365 | .2377 | .3287 | .4340 | .5049 |
| | 2 | .0373 | .1764 | .2783 | .3699 | .4711 | .5400 |
| | 3 | .0530 | .2324 | .3389 | .4318 | .5283 | .5947 |
| | 4 | .0688 | .2792 | .3984 | .4957 | .5912 | .6571 |
| | 5 | .0783 | .2982 | .4354 | .5396 | .6404 | .7094 |
| | 6 | .0819 | .3019 | .4486 | .5580 | .6662 | .7391 |
| | 7 | .0828 | .3012 | .4506 | .5615 | .6744 | .7482 |
| | 8 | .0830 | .3006 | .4498 | .5604 | .6752 | .7478 |
| | 9 | .0830 | .3004 | .4492 | .5592 | .6744 | .7454 |

**Table A.5** (Continued)

| $t$ | $\delta$ | 3 | 4 | 5 | 6 | 8 | 10 |
|---|---|---|---|---|---|---|---|
| | | | | | $b$ | | |
| 6 | 0 | .0572 | .2169 | .3564 | .4279 | .5471 | .6248 |
| | 1 | .0657 | .2359 | .3656 | .4377 | .5555 | .6308 |
| | 2 | .0911 | .2891 | .3932 | .4670 | .5799 | .6487 |
| | 3 | .1240 | .3535 | .4355 | .5129 | .6176 | .6773 |
| | 4 | .1457 | .3955 | .4828 | .5648 | .6610 | .7126 |
| | 5 | .1529 | .4082 | .5209 | .6086 | .6995 | .7470 |
| | 6 | .1539 | .4086 | .5431 | .6351 | .7250 | .7734 |
| | 7 | .1539 | .4076 | .5525 | .6466 | .7372 | .7892 |
| | 8 | .1538 | .4070 | .5555 | .6502 | .7411 | .7962 |
| | 9 | .1538 | .4070 | .5563 | .6508 | .7416 | .7985 |
| 8 | 0 | .1669 | .3793 | .5207 | .5890 | .6968 | .7369 |
| | 1 | .1830 | .3920 | .5275 | .5961 | .7021 | .7413 |
| | 2 | .2250 | .4275 | .5475 | .6166 | .7179 | .7543 |
| | 3 | .2699 | .4758 | .5779 | .6471 | .7416 | .7744 |
| | 4 | .2939 | .5197 | .6125 | .6809 | .7688 | .7988 |
| | 5 | .2982 | .5463 | .6428 | .7102 | .7938 | .8230 |
| | 6 | .2961 | .5561 | .6631 | .7299 | .8117 | .8427 |
| | 7 | .2945 | .5572 | .6736 | .7396 | .8216 | .8553 |
| | 8 | .2939 | .5564 | .6781 | .7428 | .8257 | .8616 |
| | 9 | .2938 | .5557 | .6795 | .7432 | .8269 | .8639 |
| 10 | 0 | .2686 | .4914 | .6162 | .6743 | .7672 | .8017 |
| | 1 | .2883 | .4998 | .6229 | .6815 | .7713 | .8052 |
| | 2 | .3375 | .5242 | .6421 | .7018 | .7828 | .8152 |
| | 3 | .3865 | .5602 | .6697 | .7305 | .8002 | .8308 |
| | 4 | .4104 | .5976 | .6989 | .7606 | .8202 | .8495 |
| | 5 | .4143 | .6258 | .7228 | .7845 | .8389 | .8681 |
| | 6 | .4131 | .6414 | .7375 | .7981 | .8529 | .8833 |
| | 7 | .4123 | .6480 | .7443 | .8032 | .8613 | .8932 |
| | 8 | .4119 | .6502 | .7467 | .8037 | .8652 | .8983 |
| | 9 | .4119 | .6508 | .7473 | .8029 | .8666 | .9002 |

### Table A.6 The 5% Critical Points for $\Lambda$ When $q = 1$
### (the critical points for $\delta = 9$ are approximately correct for values of $\delta > 9$)

| | | | | $b$ | | | |
|---|---|---|---|---|---|---|---|
| $t$ | $\delta$ | $3$ | $4$ | $5$ | $6$ | $8$ | $10$ |
| 3 | 0 | .0³376 | .0114 | .0411 | .0758 | .1335 | .1995 |
| | 1 | .0³559 | .0162 | .0510 | .0907 | .1557 | .2189 |
| | 2 | .00140 | .0361 | .0869 | .1446 | .2268 | .2828 |
| | 3 | .00321 | .0699 | .1469 | .2309 | .3370 | .3875 |
| | 4 | .00500 | .0937 | .1964 | .2988 | .4380 | .4947 |
| | 5 | .00581 | .0998 | .2159 | .3239 | .4874 | .5610 |
| | 6 | .00598 | .0996 | .2188 | .3272 | .4940 | .5845 |
| | 7 | .00599 | .0990 | .2184 | .3266 | .4873 | .5880 |
| | 8 | .00599 | .0989 | .2180 | .3262 | .4819 | .5872 |
| | 9 | .00599 | .0989 | .2179 | .3261 | .4797 | .5865 |
| 4 | 0 | .0260 | .1290 | .2233 | .3003 | .4218 | .4983 |
| | 1 | .0311 | .1507 | .2450 | .3185 | .4384 | .5127 |
| | 2 | .0475 | .2121 | .3089 | .3716 | .4868 | .5551 |
| | 3 | .0701 | .2829 | .3938 | .4478 | .5574 | .6180 |
| | 4 | .0858 | .3227 | .4577 | .5193 | .6274 | .6844 |
| | 5 | .0913 | .3322 | .4853 | .5634 | .6754 | .7349 |
| | 6 | .0924 | .3320 | .4921 | .5807 | .6975 | .7615 |
| | 7 | .0925 | .3313 | .4930 | .5845 | .7042 | .7707 |
| | 8 | .0925 | .3311 | .4930 | .5845 | .7052 | .7722 |
| | 9 | .0925 | .3310 | .4930 | .5842 | .7051 | .7719 |
| 5 | 0 | .1010 | .2761 | .3979 | .4905 | .5890 | .6459 |
| | 1 | .1095 | .2960 | .4146 | .5054 | .6012 | .6567 |
| | 2 | .1352 | .3504 | .4608 | .5469 | .6355 | .6872 |
| | 3 | .1720 | .4180 | .5230 | .6039 | .6839 | .7310 |
| | 4 | .2043 | .4685 | .5786 | .6578 | .7326 | .7770 |
| | 5 | .2224 | .4910 | .6121 | .6936 | .7685 | .8134 |
| | 6 | .2290 | .4967 | .6254 | .7096 | .7875 | .8343 |
| | 7 | .2308 | .4971 | .6286 | .7141 | .7945 | .8423 |
| | 8 | .2311 | .4969 | .6288 | .7146 | .7960 | .8437 |
| | 9 | .2312 | .4967 | .6286 | .7142 | .7960 | .8433 |

Table A.6 (Continued)

| $t$ | $\delta$ | $b$ | | | | | |
|---|---|---|---|---|---|---|---|
| | | 3 | 4 | 5 | 6 | 8 | 10 |
| 6 | 0 | .1740 | .3937 | .5315 | .5943 | .6912 | .7476 |
| | 1 | .1915 | .4162 | .5417 | .6042 | .6987 | .7530 |
| | 2 | .2383 | .4742 | .5705 | .6325 | .7198 | .7688 |
| | 3 | .2911 | .5380 | .6113 | .6730 | .7503 | .7922 |
| | 4 | .3232 | .5778 | .6524 | .7149 | .7830 | .8187 |
| | 5 | .3341 | .5916 | .6834 | .7476 | .8103 | .8427 |
| | 6 | .3361 | .5937 | .7010 | .7668 | .8281 | .8604 |
| | 7 | .3363 | .5934 | .7086 | .7755 | .8369 | .8707 |
| | 8 | .3363 | .5932 | .7111 | .7784 | .8402 | .8756 |
| | 9 | .3363 | .5932 | .7118 | .7792 | .8411 | .8774 |
| 8 | 0 | .3429 | .5575 | .6729 | .7278 | .8041 | .8312 |
| | 1 | .3627 | .5696 | .6794 | .7336 | .8083 | .8347 |
| | 2 | .4110 | .6019 | .6976 | .7498 | .8204 | .8447 |
| | 3 | .4594 | .6432 | .7235 | .7728 | .8377 | .8595 |
| | 4 | .4860 | .6791 | .7507 | .7972 | .8565 | .8764 |
| | 5 | .4932 | .7010 | .7730 | .8177 | .8730 | .8922 |
| | 6 | .4930 | .7101 | .7874 | .8313 | .8846 | .9047 |
| | 7 | .4921 | .7124 | .7948 | .8385 | .8912 | .9127 |
| | 8 | .4918 | .7125 | .7979 | .8415 | .8942 | .9168 |
| | 9 | .4917 | .7124 | .7990 | .8423 | .8952 | .9186 |
| 10 | 0 | .4560 | .6528 | .7491 | .7906 | .8542 | .8757 |
| | 1 | .4763 | .6609 | .7545 | .7959 | .8572 | .8783 |
| | 2 | .5242 | .6833 | .7696 | .8104 | .8655 | .8856 |
| | 3 | .5693 | .7134 | .7902 | .8302 | .8774 | .8963 |
| | 4 | .5920 | .7420 | .8110 | .8503 | .8906 | .9086 |
| | 5 | .5977 | .7621 | .8274 | .8660 | .9023 | .9203 |
| | 6 | .5978 | .7728 | .8375 | .8754 | .9110 | .9295 |
| | 7 | .5975 | .7772 | .8424 | .8795 | .9163 | .9356 |
| | 8 | .5973 | .7786 | .8442 | .8806 | .9189 | .9389 |
| | 9 | .5973 | .7790 | .8448 | .8807 | .9199 | .9403 |

### Table A.7   The 10% Critical Points for $\Lambda$ When $q = 1$
### (the critical points for $\delta = 9$ are approximately correct for values of $\delta > 9$)

| $t$ | $\delta$ | $b$ |  |  |  |  |  |
|---|---|---|---|---|---|---|---|
|  |  | *3* | *4* | *5* | *6* | *8* | *10* |
| 3 | 0 | .00256 | .0333 | .0886 | .1390 | .2183 | .2933 |
|  | 1 | .00354 | .0447 | .1062 | .1625 | .2466 | .3178 |
|  | 2 | .00741 | .0849 | .1640 | .2381 | .3301 | .3921 |
|  | 3 | .0144 | .1428 | .2473 | .3432 | .4452 | .5007 |
|  | 4 | .0202 | .1805 | .3100 | .4187 | .5422 | .6004 |
|  | 5 | .0232 | .1913 | .3351 | .4474 | .5907 | .6592 |
|  | 6 | .0238 | .1920 | .3402 | .4528 | .6023 | .6813 |
|  | 7 | .0239 | .1916 | .3403 | .4529 | .6008 | .6862 |
|  | 8 | .0239 | .1914 | .3401 | .4527 | .5982 | .6865 |
|  | 9 | .0238 | .1914 | .3400 | .4527 | .5970 | .6863 |
| 4 | 0 | .0653 | .2133 | .3278 | .4082 | .5251 | .5920 |
|  | 1 | .0759 | .2467 | .3538 | .4283 | .5424 | .6065 |
|  | 2 | .1072 | .3213 | .4254 | .4843 | .5906 | .6472 |
|  | 3 | .1462 | .4001 | .5116 | .5589 | .6558 | .7040 |
|  | 4 | .1715 | .4436 | .5720 | .6244 | .7161 | .7600 |
|  | 5 | .1806 | .4559 | .5978 | .6638 | .7556 | .8007 |
|  | 6 | .1824 | .4571 | .6048 | .6802 | .7742 | .8224 |
|  | 7 | .1827 | .4568 | .6060 | .6848 | .7805 | .8307 |
|  | 8 | .1827 | .4566 | .6061 | .6855 | .7320 | .8329 |
|  | 9 | .1827 | .4566 | .6061 | .6855 | .7822 | .8333 |
| 5 | 0 | .1822 | .3878 | .5066 | .5909 | .6775 | .7236 |
|  | 1 | .1952 | .4096 | .5234 | .6052 | .6888 | .7333 |
|  | 2 | .2324 | .4667 | .5686 | .6438 | .7196 | .7599 |
|  | 3 | .2814 | .5331 | .6260 | .6940 | .7608 | .7964 |
|  | 4 | .3214 | .5808 | .6750 | .7394 | .8001 | .8329 |
|  | 5 | .3429 | .6027 | .7042 | .7691 | .8283 | .8609 |
|  | 6 | .3607 | .6091 | .7165 | .7831 | .8434 | .8773 |
|  | 7 | .3528 | .6102 | .7201 | .7878 | .8495 | .8843 |
|  | 8 | .3532 | .6102 | .7208 | .7889 | .8513 | .8864 |
|  | 9 | .3533 | .6101 | .7208 | .7890 | .8516 | .8868 |

**Table A.7** **(Continued)**

| $t$ | $\delta$ | $b$ | | | | | |
|---|---|---|---|---|---|---|---|
| | | 3 | 4 | 5 | 6 | 8 | 10 |
| 6 | 0 | .2779 | .5052 | .6283 | .6819 | .7622 | .8061 |
| | 1 | .3000 | .5274 | .6383 | .6913 | .7689 | .8111 |
| | 2 | .3558 | .5825 | .6658 | .7173 | .7875 | .8251 |
| | 3 | .4143 | .6400 | .7028 | .7527 | .8131 | .8451 |
| | 4 | .4487 | .6754 | .7381 | .7873 | .8394 | .8667 |
| | 5 | .4608 | .6887 | .7637 | .8133 | .8607 | .8854 |
| | 6 | .4634 | .6916 | .7779 | .8283 | .8744 | .8987 |
| | 7 | .4637 | .6918 | .7842 | .8352 | .8814 | .9064 |
| | 8 | .4637 | .6917 | .7864 | .8377 | .8843 | .9102 |
| | 9 | .4637 | .6917 | .7870 | .8384 | .8853 | .9117 |
| 8 | 0 | .4624 | .6543 | .7488 | .7944 | .8533 | .8739 |
| | 1 | .4817 | .6651 | .7548 | .7993 | .8569 | .8769 |
| | 2 | .5275 | .6901 | .7711 | .8128 | .8669 | .3852 |
| | 3 | .5723 | .7283 | .7935 | .8314 | .8800 | .8971 |
| | 4 | .5977 | .7580 | .8159 | .8505 | .8954 | .9103 |
| | 5 | .6060 | .7764 | .8334 | .8663 | .9077 | .9222 |
| | 6 | .6071 | .7847 | .8445 | .8769 | .9164 | .9313 |
| | 7 | .6069 | .7874 | .8502 | .8827 | .9214 | .9372 |
| | 8 | .6067 | .7880 | .8527 | .8853 | .9237 | .9404 |
| | 9 | .6067 | .7880 | .8535 | .8862 | .9247 | .9418 |
| 10 | 0 | .5677 | .7343 | .8120 | .8443 | .8929 | .9083 |
| | 1 | .5862 | .7418 | .8166 | .8485 | .8953 | .9104 |
| | 2 | .6286 | .7619 | .8290 | .8598 | .9019 | .9163 |
| | 3 | .6678 | .7875 | .8456 | .8761 | .9112 | .9247 |
| | 4 | .6882 | .8104 | .8619 | .8902 | .9210 | .9340 |
| | 5 | .6943 | .8258 | .8744 | .9019 | .9297 | .9425 |
| | 6 | .6952 | .8337 | .8821 | .9092 | .9359 | .9492 |
| | 7 | .6951 | .8369 | .8859 | .9127 | .9398 | .9536 |
| | 8 | .6951 | .8380 | .8875 | .9141 | .9418 | .9560 |
| | 9 | .6951 | .8383 | .8880 | .9144 | .9427 | .9571 |

## Table A.8    The 1% Critical Points for $\Lambda$ When $q = 2$
(the critical points for $\delta = 9$ are approximately correct for values of $\delta > 9$)

| $t$ | $\delta$ | \multicolumn{6}{c}{$b$} |||||| |
|---|---|---|---|---|---|---|---|
| | | *3* | *4* | *5* | *6* | *8* | *10* |
| 4 | 0 | .00192 | .0262 | .0645 | .1019 | .1769 | .2315 |
| | 1 | .00245 | .0296 | .0705 | .1102 | .1863 | .2412 |
| | 2 | .00473 | .0465 | .0985 | .1447 | .2234 | .2803 |
| | 3 | .00864 | .0774 | .1510 | .2104 | .2962 | .3594 |
| | 4 | .0110 | .1003 | .1953 | .2756 | .3835 | .4580 |
| | 5 | .0109 | .1052 | .2089 | .3049 | .4426 | .5283 |
| | 6 | .0102 | .1033 | .2068 | .3066 | .4594 | .5495 |
| | 7 | .0099 | .1020 | .2039 | .3025 | .4555 | .5445 |
| | 8 | .0098 | .1016 | .2028 | .3001 | .4496 | .5367 |
| | 9 | .0098 | .1015 | .2025 | .2993 | .4465 | .5324 |
| 5 | 0 | .0143 | .0788 | .1566 | .2145 | .3375 | .4008 |
| | 1 | .0157 | .0871 | .1682 | .2278 | .3451 | .4100 |
| | 2 | .0225 | .1160 | .2072 | .2730 | .3725 | .4416 |
| | 3 | .0335 | .1611 | .2688 | .3473 | .4240 | .4998 |
| | 4 | .0413 | .1972 | .3219 | .4178 | .4892 | .5732 |
| | 5 | .0438 | .2104 | .3439 | .4514 | .5448 | .6359 |
| | 6 | .0441 | .2113 | .3457 | .4547 | .5753 | .6686 |
| | 7 | .0440 | .2104 | .3434 | .4496 | .5850 | .6758 |
| | 8 | .0440 | .2100 | .3419 | .4461 | .5855 | .6730 |
| | 9 | .0440 | .2100 | .3414 | .4448 | .5846 | .6697 |
| 6 | 0 | .0377 | .1590 | .2738 | .3294 | .4498 | .5045 |
| | 1 | .0421 | .1678 | .2934 | .3426 | .4584 | .5127 |
| | 2 | .0562 | .1955 | .3137 | .3824 | .4858 | .5383 |
| | 3 | .0757 | .2383 | .3615 | .4434 | .5318 | .5813 |
| | 4 | .0891 | .2791 | .4113 | .5058 | .5873 | .6354 |

**Table A.8** (Continued)

| $t$ | $\delta$ | b | | | | | |
|---|---|---|---|---|---|---|---|
| | | 3 | 4 | 5 | 6 | 8 | 10 |
| | 5 | .0931 | .3025 | .4445 | .5459 | .6350 | .6868 |
| | 6 | .0931 | .3098 | .4573 | .5583 | .6624 | .7220 |
| | 7 | .0927 | .3102 | .4588 | .5556 | .6718 | .7375 |
| | 8 | .0925 | .3095 | .4574 | .5503 | .6725 | .7404 |
| | 9 | .0925 | .3091 | .4563 | .5469 | .6711 | .7389 |
| 8 | 0 | .1195 | .2887 | .4194 | .5066 | .6158 | .6716 |
| | 1 | .1270 | .3027 | .4299 | .5129 | .6211 | .6758 |
| | 2 | .1494 | .3426 | .4601 | .5315 | .6372 | .6887 |
| | 3 | .1795 | .3958 | .5040 | .5615 | .6631 | .7098 |
| | 4 | .2014 | .4389 | .5474 | .5978 | .6946 | .7370 |
| | 5 | .2098 | .4577 | .5761 | .6318 | .7250 | .7656 |
| | 6 | .2113 | .4592 | .5877 | .6563 | .7474 | .7899 |
| | 7 | .2113 | .4559 | .5897 | .6691 | .7598 | .8063 |
| | 8 | .2112 | .4534 | .5889 | .6739 | .7646 | .8145 |
| | 9 | .2112 | .4523 | .5881 | .6748 | .7654 | .8172 |
| 10 | 0 | .1953 | .4212 | .5264 | .6051 | .6997 | .7472 |
| | 1 | .2055 | .4287 | .5342 | .6108 | .7045 | .7505 |
| | 2 | .2340 | .4510 | .5572 | .6277 | .7185 | .7605 |
| | 3 | .2705 | .4836 | .5912 | .6539 | .7401 | .7766 |
| | 4 | .2989 | .5166 | .6266 | .6842 | .7657 | .7971 |
| | 5 | .3118 | .5398 | .6528 | .7112 | .7892 | .8187 |
| | 6 | .3144 | .5515 | .6664 | .7299 | .8059 | .8374 |
| | 7 | .3137 | .5555 | .6711 | .7393 | .8146 | .8504 |
| | 8 | .3131 | .5564 | .6720 | .7426 | .8176 | .8576 |
| | 9 | .3129 | .5566 | .6719 | .7432 | .8180 | .8605 |

### Table A.9    The 5% Critical Points for $\Lambda$ When $q = 2$
### (the critical points for $\delta = 9$ are approximately correct for values of $\delta > 9$)

| | | | | $b$ | | | |
|---|---|---|---|---|---|---|---|
| $t$ | $\delta$ | 3 | 4 | 5 | 6 | 8 | 10 |
| 4 | 0 | .0136 | .0765 | .1425 | .1979 | .2912 | .3496 |
| | 1 | .0165 | .0852 | .1558 | .2129 | .3059 | .3641 |
| | 2 | .0270 | .1232 | .2044 | .2654 | .3552 | .4137 |
| | 3 | .0420 | .1789 | .2793 | .3484 | .4369 | .4976 |
| | 4 | .0508 | .2167 | .3366 | .4210 | .5225 | .5883 |
| | 5 | .0517 | .2270 | .3565 | .4544 | .5774 | .6490 |
| | 6 | .0504 | .2264 | .3575 | .4603 | .5962 | .6705 |
| | 7 | .0495 | .2251 | .3556 | .4584 | .5969 | .6711 |
| | 8 | .0493 | .2246 | .3547 | .4569 | .5941 | .6677 |
| | 9 | .0492 | .2246 | .3545 | .4564 | .5924 | .6655 |
| 5 | 0 | .0640 | .1737 | .2754 | .3399 | .4652 | .5235 |
| | 1 | .0595 | .1887 | .2923 | .3576 | .4756 | .5346 |
| | 2 | .0794 | .2349 | .3436 | .4114 | .5084 | .5693 |
| | 3 | .1067 | .2973 | .4143 | .4887 | .5616 | .6250 |
| | 4 | .1244 | .3428 | .4705 | .5557 | .6213 | .6877 |
| | 5 | .1301 | .3604 | .4950 | .5887 | .6684 | .7374 |
| | 6 | .1310 | .3634 | .4997 | .5956 | .6940 | .7637 |
| | 7 | .1310 | .3630 | .4989 | .5940 | .7032 | .7715 |
| | 8 | .1310 | .3627 | .4981 | .5922 | .7050 | .7717 |
| | 9 | .1310 | .3626 | .4979 | .5915 | .7048 | .7704 |
| 6 | 0 | .1075 | .2833 | .4078 | .4665 | .5748 | .6250 |
| | 1 | .1174 | .2959 | .4197 | .4805 | .5842 | .6334 |
| | 2 | .1461 | .3329 | .4545 | .5206 | .6121 | .6580 |
| | 3 | .1817 | .3642 | .5045 | .5771 | .6547 | .6961 |
| | 4 | .2052 | .4296 | .5528 | .6313 | .7014 | .7400 |

**Table A.9** (Continued)

| | | | | $b$ | | | |
|---|---|---|---|---|---|---|---|
| $t$ | $\delta$ | 3 | 4 | 5 | 6 | 8 | 10 |
| | 5 | .2131 | .4554 | .5844 | .6662 | .7393 | .7790 |
| | 6 | .2141 | .4645 | .5978 | .6795 | .7612 | .8052 |
| | 7 | .2138 | .4661 | .6009 | .6806 | .7697 | .8175 |
| | 8 | .2137 | .4660 | .6008 | .6785 | .7715 | .8211 |
| | 9 | .2137 | .4658 | .6004 | .6769 | .7713 | .8211 |
| 8 | 0 | .2350 | .4300 | .5549 | .6295 | .7178 | .7619 |
| | 1 | .2470 | .4451 | .5653 | .6357 | .7229 | .7658 |
| | 2 | .2805 | .4453 | .5942 | .6537 | .7375 | .7773 |
| | 3 | .3211 | .5367 | .6335 | .6807 | .7596 | .7952 |
| | 4 | .3492 | .5769 | .6704 | .7113 | .7850 | .8168 |
| | 5 | .3606 | .5960 | .6946 | .7387 | .8081 | .8382 |
| | 6 | .3632 | .6002 | .7054 | .7580 | .8248 | .8558 |
| | 7 | .3635 | .5993 | .7083 | .7686 | .8343 | .8676 |
| | 8 | .3635 | .5981 | .7084 | .7730 | .8384 | .8739 |
| | 9 | .3635 | .5975 | .7082 | .7743 | .8396 | .8764 |
| 10 | 0 | .3330 | .5576 | .6483 | .7129 | .7853 | .8214 |
| | 1 | .3456 | .5656 | .6558 | .7182 | .7895 | .8243 |
| | 2 | .3790 | .5581 | .6769 | .7331 | .8012 | .8328 |
| | 3 | .4193 | .6188 | .7061 | .7549 | .8186 | .8458 |
| | 4 | .4499 | .6479 | .7047 | .7788 | .8379 | .8614 |
| | 5 | .4646 | .5677 | .7554 | .7994 | .8551 | .8770 |
| | 6 | .4686 | .6777 | .7662 | .8134 | .8672 | .8899 |
| | 7 | .4689 | .6814 | .7704 | .8207 | .8738 | .8990 |
| | 8 | .4686 | .6924 | .7715 | .8237 | .8765 | .9040 |
| | 9 | .4684 | .6826 | .7717 | .8245 | .8772 | .9063 |

### Table A.10 The 10% Critical Points for $\Lambda$ When $q = 2$
(the critical points for $\delta = 9$ are approximately correct for values of $\delta > 9$)

| $t$ | $\delta$ | $b$ | | | | | |
|---|---|---|---|---|---|---|---|
| | | *3* | *4* | *5* | *6* | *8* | *10* |
| 4 | 0 | .0318 | .1230 | .2030 | .2660 | .3641 | .4209 |
| | 1 | .0378 | .1379 | .2213 | .2854 | .3816 | .4380 |
| | 2 | .0574 | .1888 | .2818 | .3467 | .4360 | .4917 |
| | 3 | .0832 | .2578 | .3656 | .4345 | .5183 | .5743 |
| | 4 | .0985 | .3029 | .4266 | .5067 | .5984 | .6568 |
| | 5 | .1014 | .3167 | .4494 | .5406 | .6487 | .7105 |
| | 6 | .1001 | .3176 | .4526 | .5486 | .6678 | .7313 |
| | 7 | .0991 | .3167 | .4518 | .5485 | .6709 | .7346 |
| | 8 | .0987 | .3164 | .4512 | .5477 | .6699 | .7334 |
| | 9 | .0987 | .3163 | .4511 | .5473 | .6690 | .7323 |
| 5 | 0 | .0966 | .2460 | .3537 | .4173 | .5370 | .5900 |
| | 1 | .1063 | .2650 | .3732 | .4367 | .5486 | .6019 |
| | 2 | .1373 | .3196 | .4290 | .4927 | .5833 | .6371 |
| | 3 | .1761 | .3878 | .5005 | .5677 | .6352 | .6895 |
| | 4 | .2001 | .4357 | .5550 | .6296 | .6895 | .7447 |
| | 5 | .2079 | .4548 | .5796 | .6609 | .7305 | .7868 |
| | 6 | .2094 | .4589 | .5858 | .6695 | .7529 | .8094 |
| | 7 | .2095 | .4591 | .5861 | .6698 | .7615 | .8173 |
| | 8 | .2095 | .4590 | .5857 | .6690 | .7637 | .8186 |
| | 9 | .2095 | .4589 | .5855 | .6686 | .7640 | .8183 |
| 6 | 0 | .1699 | .3655 | .4866 | .5440 | .6409 | .6871 |
| | 1 | .1836 | .3798 | .4993 | .5578 | .6504 | .6953 |
| | 2 | .2215 | .4201 | .5350 | .5961 | .6776 | .7186 |
| | 3 | .2657 | .4730 | .5837 | .6478 | .7169 | .7530 |
| | 4 | .2942 | .5180 | .6289 | .6958 | .7578 | .7907 |

**Table A.10**   (Continued)

| t | δ | 3 | 4 | 5 | 6 | 8 | 10 |
|---|---|---|---|---|---|---|---|
| | | | | | *b* | | |
| | 5 | .3045 | .5436 | .6581 | .7268 | .7899 | .8229 |
| | 6 | .3064 | .5533 | .6713 | .7400 | .8085 | .8443 |
| | 7 | .3064 | .5556 | .6752 | .7430 | .8164 | .8549 |
| | 8 | .3063 | .5558 | .6758 | .7426 | .8187 | .8586 |
| | 9 | .3063 | .6558 | .6757 | .7418 | .8190 | .8594 |
| 8 | 0 | .3162 | .5123 | .6275 | .6926 | .7680 | .8055 |
| | 1 | .3306 | .5271 | .6374 | .6986 | .7728 | .8092 |
| | 2 | .3691 | .5660 | .6645 | .7156 | .7864 | .8198 |
| | 3 | .4133 | .6132 | .7000 | .7402 | .8061 | .8357 |
| | 4 | .4432 | .6499 | .7322 | .7671 | .8279 | .8541 |
| | 5 | .4555 | .6684 | .7534 | .7905 | .8471 | .8718 |
| | 6 | .4587 | .6739 | .7633 | .8068 | .8609 | .8861 |
| | 7 | .4592 | .6743 | .7656 | .8160 | .8688 | .8956 |
| | 8 | .4592 | .6738 | .7672 | .8201 | .8725 | .9009 |
| | 9 | .4592 | .6735 | .7672 | .8216 | .8739 | .9032 |
| 10 | 0 | .4206 | .6306 | .7103 | .7660 | .8261 | .8562 |
| | 1 | .4337 | .6386 | .7174 | .7709 | .8299 | .8589 |
| | 2 | .4676 | .6603 | .7368 | .7843 | .8403 | .8664 |
| | 3 | .5074 | .6886 | .7626 | .8035 | .8553 | .8777 |
| | 4 | .5373 | .7146 | .7872 | .8237 | .8714 | .8908 |
| | 5 | .5521 | .7320 | .8046 | .8408 | .8854 | .9034 |
| | 6 | .5568 | .7407 | .8139 | .8524 | .8952 | .9137 |
| | 7 | .5577 | .7441 | .8177 | .8586 | .9007 | .9208 |
| | 8 | .5576 | .7451 | .8189 | .8614 | .9032 | .9249 |
| | 9 | .5575 | .7454 | .8191 | .8623 | .9040 | .9269 |

## Table A.11   Approximate Critical Points of $\Lambda$ When $q = 1$ and $r = 1$

|   |   | 3 | | | 4 | | | 5 | | | 6 | | | 8 | | 1 |
|---|---|---|---|---|---|---|---|---|---|---|---|---|---|---|---|---|
| $t$ | $\delta$ | 1% | 5% | 10% | 1% | 5% | 10% | 1% | 5% | 10% | 1% | 5% | 10% | 1% | 5% | |
| 3 | 0 | $.0^4135$ | $.0^3499$ | $.0^2236$ | | | | | | | | | | | | |
|   | 1 | $.0^4171$ | $.0^3615$ | $.0^2288$ | | | | | | | | | | | | |
|   | 2 | $.0^4454$ | $.0^2128$ | $.0^2538$ | | | | | | | | | | | | |
|   | 3 | $.0^3112$ | $.0^2245$ | $.0^2929$ | | | | | | | | | | | | |
|   | 4 | $.0^3150$ | $.0^2310$ | .0115 | | | | | | | | | | | | |
|   | 5 | $.0^3142$ | $.0^2308$ | .0116 | | | | | | | | | | | | |
|   | 6 | $.0^3130$ | $.0^2293$ | .0112 | | | | | | | | | | | | |
|   | 7 | $.0^3124$ | $.0^2287$ | .0111 | | | | | | | | | | | | |
|   | 8 | $.0^3123$ | $.0^2285$ | .0110 | | | | | | | | | | | | |
|   | 9 | $.0^3123$ | $.0^2284$ | .0110 | | | | | | | | | | | | |
|   | 10 | $.0^3123$ | $.0^2284$ | .0110 | | | | | | | | | | | | |
| 4 | 0 | $.0^2173$ | .0126 | .0298 | .0255 | .0791 | .1300 | | | | | | | | | |
|   | 1 | $.0^2210$ | .0150 | .0351 | .0299 | .0903 | .1465 | | | | | | | | | |
|   | 2 | $.0^2407$ | .0248 | .0542 | .0472 | .1276 | .1969 | | | | | | | | | |
|   | 3 | $.0^2778$ | .0398 | .0805 | .0750 | .1780 | .2592 | | | | | | | | | |
|   | 4 | .0102 | .0490 | .0964 | .0940 | .2100 | .2975 | | | | | | | | | |
|   | 5 | .0105 | .0509 | .1003 | .0982 | .2187 | .3091 | | | | | | | | | |
|   | 6 | .0103 | .0504 | .1001 | .0972 | .2187 | .3101 | | | | | | | | | |
|   | 7 | .0102 | .0501 | .0997 | .0964 | .2179 | .3097 | | | | | | | | | |
|   | 8 | .0101 | .0500 | .0996 | .0962 | .2177 | .3095 | | | | | | | | | |
|   | 9 | .0101 | .0500 | .0996 | .0962 | .2176 | .3094 | | | | | | | | | |
|   | 10 | .0101 | .0500 | .0996 | .0962 | .2176 | .3094 | | | | | | | | | |
| 5 | 0 | $.0^2819$ | .0357 | .0679 | .0840 | .1796 | .2512 | .1802 | .3061 | .3869 | | | | | | |
|   | 1 | .0115 | .0464 | .0854 | .0889 | .1896 | .2646 | .1910 | .3210 | .4036 | | | | | | |
|   | 2 | .0256 | .0832 | .1391 | .1079 | .2224 | .3052 | .2254 | .3645 | .4501 | | | | | | |
|   | 3 | .0438 | .1233 | .1933 | .1422 | .2732 | .3629 | .2776 | .4237 | .5097 | | | | | | |
|   | 4 | .0479 | .1350 | .2113 | .1783 | .3208 | .4138 | .3240 | .4729 | .5574 | | | | | | |

| Group | | C1 | C2 | C3 | C4 | C5 | C6 | C7 | C8 | C9 | C10 | C11 | C12 | C13 | C14 | |
|---|---|---|---|---|---|---|---|---|---|---|---|---|---|---|---|---|
| | 5 | .0447 | .1316 | .2097 | .2001 | .3483 | .4426 | .3464 | .4973 | .5816 | | | | | | |
| | 6 | .0427 | .1288 | .2071 | .2075 | .3582 | .4533 | .3505 | .5037 | .5891 | | | | | | |
| | 7 | .0422 | .1280 | .2063 | .2086 | .3603 | .4560 | .3490 | .5038 | .5901 | | | | | | |
| | 8 | .0421 | .1279 | .2062 | .2084 | .3604 | .4563 | .3477 | .5032 | .5900 | | | | | | |
| | 9 | .0421 | .1278 | .2062 | .2083 | .3603 | .4563 | .3472 | .5029 | .5898 | | | | | | |
| | 10 | .0421 | .1278 | .2062 | .2082 | .3603 | .4563 | .3471 | .5028 | .5898 | | | | | | |
| 6 | 0 | .0267 | .0798 | .1292 | .1309 | .2462 | .3254 | .2588 | .3929 | .4727 | .3476 | .4832 | .5590 | | | |
| | 1 | .0313 | .0911 | .1455 | .1436 | .2646 | .3462 | .2709 | .4078 | .4886 | .3593 | .4962 | .5721 | | | |
| | 2 | .0493 | .1288 | .1960 | .1845 | .3182 | .4040 | .3090 | .4513 | .5328 | .3950 | .5333 | .6085 | | | |
| | 3 | .0783 | .1805 | .2597 | .2450 | .3884 | .4750 | .3675 | .5110 | .5902 | .4498 | .5854 | .6569 | | | |
| | 4 | .0976 | .2134 | .2996 | .2929 | .4405 | .5263 | .4223 | .5629 | .6379 | .5048 | .6340 | .7002 | | | |
| | 5 | .1001 | .2207 | .3107 | .3097 | .4609 | .5476 | .4517 | .5908 | .6639 | .5393 | .6642 | .7271 | | | |
| | 6 | .0972 | .2188 | .3104 | .3089 | .4632 | .5519 | .4592 | .5995 | .6729 | .5511 | .6759 | .7383 | | | |
| | 7 | .0954 | .2170 | .3091 | .3057 | .4617 | .5515 | .4585 | .6004 | .6746 | .5516 | .6781 | .7413 | | | |
| | 8 | .0949 | .2164 | .3086 | .3041 | .4607 | .5509 | .4571 | .5999 | .6746 | .5498 | .6777 | .7416 | | | |
| | 9 | .0948 | .2163 | .3085 | .3036 | .4603 | .5507 | .4565 | .5996 | .6744 | .5487 | .6772 | .7414 | | | |
| | 10 | .0947 | .2163 | .3085 | .3035 | .4603 | .5507 | .4563 | .5995 | .6744 | .5483 | .6770 | .7413 | | | |
| 8 | 0 | .0567 | .1369 | .2020 | .2396 | .3731 | .4539 | .3840 | .5117 | .5908 | .4869 | .6114 | .6760 | .6249 | .7256 | .7 |
| | 1 | .0637 | .1508 | .2202 | .2512 | .3878 | .4696 | .3950 | .5301 | .6035 | .4959 | .6204 | .6846 | .6299 | .7302 | .7 |
| | 2 | .0904 | .1967 | .2763 | .2883 | .4310 | .5140 | .4297 | .5662 | .6389 | .5231 | .6458 | .7083 | .6448 | .7437 | .7 |
| | 3 | .1394 | .2672 | .3549 | .3478 | .4929 | .5739 | .4839 | .6170 | .6858 | .5651 | .6825 | .7411 | .6688 | .7640 | .8 |
| | 4 | .1878 | .3285 | .4190 | .4082 | .5503 | .6268 | .5387 | .6642 | .7274 | .6113 | .7203 | .7738 | .6982 | .7876 | .8 |
| | 5 | .2109 | .3576 | .4496 | .4450 | .5846 | .6583 | .5734 | .6933 | .7528 | .6473 | .7490 | .7981 | .7271 | .8097 | .8 |
| | 6 | .2142 | .3640 | .4578 | .4566 | .5970 | .6705 | .5861 | .7049 | .7633 | .6661 | .7644 | .8115 | .7495 | .8265 | .8 |
| | 7 | .2125 | .3635 | .4583 | .4567 | .5990 | .6734 | .5879 | .7075 | .7663 | .6714 | .7699 | .8169 | .7626 | .8366 | .8 |
| | 8 | .2112 | .3626 | .4579 | .4550 | .5984 | .6735 | .5872 | .7076 | .7667 | .6710 | .7708 | .8183 | .7680 | .8413 | .8 |
| | 9 | .2108 | .3623 | .4576 | .4540 | .5980 | .6733 | .5866 | .7074 | .7667 | .6697 | .7705 | .8185 | .7692 | .8428 | .8 |
| | 10 | .2107 | .3623 | .4576 | .4537 | .5978 | .6732 | .5864 | .7073 | .7667 | .6689 | .7702 | .8184 | .7688 | .8431 | .8 |

## Table A.12  Approximate Critical Points of Λ When $q = 1$ and $r = 2$

| t | δ | b=4 1% | b=4 5% | b=4 10% | b=5 1% | b=5 5% | b=5 10% | b=6 1% | b=6 5% | b=6 10% | b=8 1% | b=8 5% | b=8 10% |
|---|---|---|---|---|---|---|---|---|---|---|---|---|---|
| 3 | 0 | $.0^2220$ | .0123 | .0261 | .0113 | .0385 | .0664 | .0270 | .0718 | .1110 | .0690 | .1399 | .1926 |
|   | 1 | $.0^2228$ | .0134 | .0290 | .0123 | .0424 | .0732 | .0289 | .0774 | .1200 | .0707 | .1458 | .2019 |
|   | 2 | $.0^2377$ | .0204 | .0426 | .0181 | .0583 | .0976 | .0389 | .0996 | .1512 | .0835 | .1709 | .2353 |
|   | 3 | $.0^2617$ | .0299 | .0595 | .0280 | .0815 | .1306 | .0570 | .1338 | .1953 | .1116 | .2148 | .2874 |
|   | 4 | $.0^2697$ | .0336 | .0666 | .0343 | .0962 | .1515 | .0715 | .1596 | .2276 | .1417 | .2566 | .3338 |
|   | 5 | $.0^2660$ | .0330 | .0665 | .0347 | .0990 | .1567 | .0755 | .1683 | .2396 | .1586 | .2794 | .3590 |
|   | 6 | $.0^2627$ | .0322 | .0655 | .0333 | .0974 | .1558 | .0743 | .1683 | .2409 | .1633 | .2866 | .3674 |
|   | 7 | $.0^2617$ | .0319 | .0651 | .0325 | .0962 | .1547 | .0730 | .1670 | .2400 | .1634 | .2876 | .3690 |
|   | 8 | $.0^2615$ | .0318 | .0650 | .0322 | .0958 | .1542 | .0725 | .1663 | .2395 | .1630 | .2874 | .3690 |
|   | 9 | $.0^2614$ | .0318 | .0650 | .0322 | .0957 | .1541 | .0723 | .1662 | .2393 | .1629 | .2873 | .3689 |
|   | 10 | $.0^2614$ | .0318 | .0650 | .0322 | .0957 | .1541 | .0723 | .1661 | .2393 | .1628 | .2872 | .3689 |
| 4 | 0 | .0314 | .0771 | .1156 | .0699 | .1420 | .1954 | .1374 | .2310 | .2925 | .2265 | .3400 | .4085 |
|   | 1 | .0319 | .0806 | .1220 | .0770 | .1549 | .2121 | .1411 | .2389 | .3032 | .2337 | .3501 | .4202 |
|   | 2 | .0388 | .0978 | .1475 | .1038 | .1967 | .2620 | .1588 | .2670 | .3373 | .2585 | .3815 | .4543 |
|   | 3 | .0541 | .1279 | .1872 | .1435 | .2516 | .3234 | .1930 | .3126 | .3877 | .3006 | .4288 | .5026 |
|   | 4 | .0679 | .1532 | .2195 | .1685 | .2859 | .3617 | .2292 | .3561 | .4334 | .3456 | .4753 | .5478 |
|   | 5 | .0726 | .1633 | .2333 | .1711 | .2933 | .3723 | .2506 | .3815 | .4598 | .3754 | .5054 | .5768 |
|   | 6 | .0719 | .1640 | .2357 | .1665 | .2905 | .3715 | .2569 | .3902 | .4696 | .3867 | .5179 | .5895 |
|   | 7 | .0704 | .1627 | .2349 | .1634 | .2880 | .3698 | .2570 | .3915 | .4717 | .3882 | .5207 | .5930 |
|   | 8 | .0697 | .1619 | .2343 | .1623 | .2871 | .3691 | .2562 | .3912 | .4717 | .3873 | .5207 | .5935 |
|   | 9 | .0695 | .1616 | .2340 | .1621 | .2868 | .3689 | .2558 | .3910 | .4716 | .3865 | .5203 | .5934 |
|   | 10 | .0695 | .1616 | .2340 | .1620 | .2868 | .3689 | .2557 | .3909 | .4716 | .3862 | .5201 | .5933 |
| 5 | 0 | .0732 | .1413 | .1905 | .1479 | .2417 | .3024 | .2337 | .3443 | .4107 | .3515 | .4688 | .5341 |
|   | 1 | .0750 | .1481 | .2014 | .1521 | .2517 | .3162 | .2393 | .3532 | .4213 | .3605 | .4790 | .5447 |
|   | 2 | .0918 | .1792 | .2419 | .1771 | .2902 | .3622 | .2607 | .3819 | .4535 | .3889 | .5093 | .5751 |
|   | 3 | .1265 | .2309 | .3021 | .2261 | .3512 | .4275 | .3002 | .4276 | .5009 | .4344 | .5537 | .6175 |
|   | 4 | .1569 | .2727 | .3487 | .2708 | .4015 | .4784 | .3454 | .4749 | .5473 | .4833 | .5984 | .6585 |

| | | | | | | | | | | | | |
|---|---|---|---|---|---|---|---|---|---|---|---|---|
| 5 | .1673 | .2887 | .3677 | .2902 | .4238 | .5014 | .3778 | .5074 | .5785 | .5179 | .6294 | .6868 |
| 6 | .1662 | .2901 | .3709 | .2932 | .4286 | .5071 | .3917 | .5221 | .5931 | .5322 | .6436 | .7004 |
| 7 | .1640 | .2885 | .3702 | .2922 | .4286 | .5077 | .3946 | .5262 | .5977 | .5340 | .6470 | .7046 |
| 8 | .1629 | .2877 | .3696 | .2916 | .4282 | .5075 | .3941 | .5265 | .5985 | .5320 | .6467 | .7051 |
| 9 | .1627 | .2875 | .3694 | .2914 | .4281 | .5074 | .3935 | .5263 | .5985 | .5303 | .6459 | .7048 |
| 10 | .1626 | .2874 | .3694 | .2914 | .4281 | .5074 | .3932 | .5261 | .5985 | .5295 | .6455 | .7046 |
| 6  0 | .1205 | .2031 | .2580 | .2120 | .3207 | .3871 | .3040 | .4183 | .4837 | .4605 | .5655 | .6211 |
| 1 | .1229 | .2112 | .2701 | .2225 | .3342 | .4020 | .3115 | .4287 | .4955 | .4654 | .5720 | .6283 |
| 2 | .1444 | .2470 | .3145 | .2570 | .3750 | .4448 | .3387 | .4615 | .5304 | .4828 | .5923 | .6496 |
| 3 | .1903 | .3076 | .3813 | .3123 | .4343 | .5038 | .3872 | .5121 | .5804 | .5150 | .6249 | .6816 |
| 4 | .2338 | .3596 | .4357 | .3660 | .4884 | .5560 | .4405 | .5627 | .6277 | .5554 | .6619 | .7158 |
| 5 | .2516 | .3822 | .4603 | .3938 | .5178 | .5854 | .4768 | .5959 | .6581 | .5909 | .6925 | .7432 |
| 6 | .2522 | .3858 | .4657 | .3963 | .5246 | .5944 | .4911 | .6098 | .6714 | .6124 | .7109 | .7597 |
| 7 | .2495 | .3845 | .4655 | .3898 | .5221 | .5943 | .4935 | .6132 | .6752 | .6209 | .7188 | .7670 |
| 8 | .2481 | .3835 | .4649 | .3843 | .5190 | .5926 | .4927 | .6133 | .6757 | .6226 | .7210 | .7695 |
| 9 | .2476 | .3832 | .4647 | .3816 | .5173 | .5916 | .4920 | .6130 | .6757 | .6223 | .7213 | .7700 |
| 10 | .2476 | .3831 | .4646 | .3807 | .5167 | .5912 | .4917 | .6128 | .6756 | .6219 | .7212 | .7700 |
| 8  0 | .1942 | .2916 | .3514 | .3300 | .4402 | .5021 | .4220 | .5322 | .5915 | .5600 | .6575 | .7071 |
| 1 | .1970 | .2996 | .3628 | .3358 | .4490 | .5126 | .4293 | .5410 | .6009 | .5657 | .6636 | .7132 |
| 2 | .2183 | .3327 | .4025 | .3592 | .4784 | .5445 | .4540 | .5681 | .6285 | .5838 | .6815 | .7307 |
| 3 | .2689 | .3938 | .4673 | .4051 | .5273 | .5934 | .4978 | .6106 | .6692 | .6145 | .7094 | .7565 |
| 4 | .3318 | .4601 | .5324 | .4621 | .5813 | .6441 | .5504 | .6571 | .7113 | .6526 | .7414 | .7848 |
| 5 | .3755 | .5039 | .5744 | .5073 | .6218 | .6810 | .5933 | .6932 | .7431 | .6879 | .7695 | .8090 |
| 6 | .3908 | .5207 | .5915 | .5291 | .6420 | .6997 | .6155 | .7124 | .7604 | .7117 | .7883 | .8250 |
| 7 | .3913 | .5234 | .5954 | .5341 | .6480 | .7060 | .6216 | .7189 | .7669 | .7226 | .7974 | .8332 |
| 8 | .3890 | .5226 | .5954 | .5331 | .6485 | .7073 | .6211 | .7197 | .7684 | .7254 | .8004 | .8362 |
| 9 | .3876 | .5218 | .5951 | .5317 | .6480 | .7072 | .6195 | .7193 | .7684 | .7250 | .8009 | .8370 |
| 10 | .3871 | .5215 | .5949 | .5310 | .6476 | .7070 | .6186 | .7188 | .7682 | .7242 | .8007 | .8370 |

*Note:* This table could also be used to obtain percentage points of $\Lambda$ when $q = 2$ and $r = 1$ by interchanging the values of $b$ and $t$ and reading the appropriate value for $q = 1$ and $r = 2$.

# Table A.13   Approximate Critical Points of Λ When q = 2 and r = 2

| | | b | | | | | | | | | | | | |
|---|---|---|---|---|---|---|---|---|---|---|---|---|---|---|
| | | 4 | | | 5 | | | 6 | | | 8 | | | |
| t | δ | 1% | 5% | 10% | 1% | 5% | 10% | 1% | 5% | 10% | 1% | 5% | 10% | |
| 4 | 0 | .0353 | .0754 | .1067 | | | | | | | | | | |
| | 1 | .0337 | .0765 | .1110 | | | | | | | | | | |
| | 2 | .0401 | .0929 | .1356 | | | | | | | | | | |
| | 3 | .0539 | .1193 | .1704 | | | | | | | | | | |
| | 4 | .0613 | .1342 | .1904 | | | | | | | | | | |
| | 5 | .0606 | .1356 | .1941 | | | | | | | | | | |
| | 6 | .0585 | .1335 | .1927 | | | | | | | | | | |
| | 7 | .0574 | .1323 | .1917 | | | | | | | | | | |
| | 8 | .0572 | .1320 | .1913 | | | | | | | | | | |
| | 9 | .0571 | .1319 | .1912 | | | | | | | | | | |
| | 10 | .0571 | .1319 | .1912 | | | | | | | | | | |
| 5 | 0 | .0735 | .1334 | .1755 | .1479 | .2280 | .2787 | | | | | | | |
| | 1 | .0738 | .1380 | .1839 | .1490 | .2338 | .2879 | | | | | | | |
| | 2 | .0871 | .1636 | .2178 | .1648 | .2605 | .3212 | | | | | | | |
| | 3 | .1149 | .2054 | .2672 | .2005 | .3071 | .3728 | | | | | | | |
| | 4 | .1359 | .2358 | .3022 | .2365 | .3499 | .4179 | | | | | | | |
| | 5 | .1405 | .2449 | .3142 | .2541 | .3716 | .4412 | | | | | | | |
| | 6 | .1382 | .2441 | .3149 | .2566 | .3767 | .4479 | | | | | | | |
| | 7 | .1361 | .2424 | .3138 | .2547 | .3762 | .4484 | | | | | | | |
| | 8 | .1352 | .2417 | .3132 | .2533 | .3753 | .4479 | | | | | | | |
| | 9 | .1351 | .2415 | .3131 | .2528 | .3750 | .4476 | | | | | | | |
| | 10 | .1350 | .2415 | .3131 | .2526 | .3749 | .4475 | | | | | | | |

| | | | | | | | | | | | | | |
|---|---|---|---|---|---|---|---|---|---|---|---|---|---|
| 6 | 0 | .1125 | .1855 | .2337 | .2161 | .3070 | .3614 | .2860 | .3826 | .4380 | .5260 | .6170 | .6642 |
| | 1 | .1159 | .1937 | .2453 | .2200 | .3150 | .3720 | .2913 | .3913 | .4487 | .5315 | .6231 | .6705 |
| | 2 | .1373 | .2271 | .2858 | .2406 | .3445 | .4061 | .3147 | .4217 | .4823 | .5497 | .6418 | .6891 |
| | 3 | .1784 | .2804 | .3444 | .2813 | .3925 | .4569 | .3600 | .4710 | .5325 | .5812 | .6714 | .7171 |
| | 4 | .2131 | .3229 | .3899 | .3227 | .4375 | .5025 | .4094 | .5201 | .5799 | .6207 | .7057 | .7482 |
| | 5 | .2239 | .3387 | .4084 | .3442 | .4619 | .5277 | .4399 | .5502 | .6091 | .6570 | .7358 | .7748 |
| | 6 | .2208 | .3388 | .4107 | .3475 | .4683 | .5357 | .4485 | .5607 | .6203 | .6802 | .7553 | .7921 |
| | 7 | .2165 | .3359 | .4091 | .3446 | .4676 | .5364 | .4471 | .5616 | .6224 | .6893 | .7639 | .8004 |
| | 8 | .2145 | .3344 | .4080 | .3422 | .4662 | .5357 | .4445 | .5604 | .6220 | .6900 | .7660 | .8030 |
| | 9 | .2139 | .3339 | .4076 | .3411 | .4655 | .5352 | .4431 | .5596 | .6216 | .6880 | .7656 | .8033 |
| | 10 | .2137 | .3338 | .4076 | .3408 | .4653 | .5351 | .4426 | .5593 | .6214 | .6863 | .7648 | .8030 |
| 8 | 0 | .1878 | .2748 | .3280 | .3209 | .4180 | .4726 | .4074 | .5042 | .5567 | | | |
| | 1 | .1920 | .2836 | .3396 | .3253 | .4256 | .4820 | .4119 | .5112 | .5649 | | | |
| | 2 | .2159 | .3175 | .3789 | .3453 | .4517 | .5110 | .4306 | .5345 | .5902 | | | |
| | 3 | .2654 | .3752 | .4395 | .3859 | .4958 | .5560 | .4682 | .5736 | .6292 | | | |
| | 4 | .3183 | .4314 | .4955 | .4354 | .5442 | .6024 | .5152 | .6173 | .6702 | | | |
| | 5 | .3468 | .4624 | .5269 | .4727 | .5794 | .6355 | .5532 | .6509 | .7008 | | | |
| | 6 | .3508 | .4700 | .5365 | .4887 | .5956 | .6515 | .5724 | .6684 | .7170 | | | |
| | 7 | .3461 | .4684 | .5367 | .4909 | .5997 | .6564 | .5775 | .6741 | .7229 | | | |
| | 8 | .3422 | .4661 | .5354 | .4888 | .5993 | .6570 | .5772 | .6749 | .7242 | | | |
| | 9 | .3405 | .4649 | .5347 | .4871 | .5985 | .6566 | .5761 | .6746 | .7243 | | | |
| | 10 | .3400 | .4645 | .5344 | .4862 | .5980 | .6563 | .5754 | .6742 | .7241 | | | |

# References

Black, A.L. 1970. Adventitious Roots, Tillers, and Grain Yields of Spring Wheat as Influenced by N-P Fertilization. *Agronomy Journal* 62:32–36.

Box, G.E.P., Hunter, W.G., and Hunter, J.G. 1978. *Statistics for Experimenters*. New York: John Wiley.

Cochran, W.G., and Cox, G.M. 1957. *Experimental Design*, 2nd ed. New York: John Wiley.

Daniel, Cuthbert. 1959. Use of Half-Normal Plots in Interpreting Factorial Two-Level Experiments. *Technometrics* 1:311–341.

Davies, Owen L. (ed.). 1954. *The Design and Analysis of Industrial Experiments*. New York: Hafner.

Gallant, A. Ronald. 1987. *Nonlinear Statistical Models*. New York: John Wiley.

Graybill, F.A. 1961. *An Introduction to Linear Statistical Models. Volume I*. New York: McGraw-Hill.

Graybill, F.A. 1976. *Theory and Application of the Linear Model*. North Scituate, Mass.: Duxbury.

Hegemann, V.J., and Johnson, D.E. 1976. On Analyzing Two-Way AOV Data with Interaction. *Technometrics* 18:273–281.

Hicks, C. R. 1982. *Fundamental Concepts in the Design of Experiments*. New York: Holt, Rinehart and Winston.

Johnson, D.E., and Graybill, F.A. 1972. An Analysis of a Two-Way Model with Interaction and No Replication. *Journal of the American Statistical Association* 67:862–868.

Kempthorne, O. 1952. *The Design and Analysis of Experiments*. New York: John Wiley.

Mandel, John. 1961. Non-additivity in Two-Way Analysis of Variance. *Journal of the American Statistical Association* 56:878–888.

Mandel, John. 1971. A New Analysis of Variance Model for Non-additive Data. *Technometrics* 13:1–18.

Marasinghe, M.G., and Johnson, D.E. 1981. Testing Subhypotheses in the Multiplicative Interaction Model. *Technometrics* 23:385–393.

Marasinghe, M.G., and Johnson, D.E. 1982a. A Test of Incomplete Additivity in the Multiplicative Interaction Model. *Journal of the American Statistical Association* 77:869–877.

Marasinghe, M.G., and Johnson, D.E. 1982b. Estimating $\sigma^2$ in the Multiplicative Interaction Model. *Communications in Statistics Theory and Methods* 11:315–324.

Milliken, G.A., and Graybill, F.A. 1970. Extensions of the General Linear Hypothesis Model. *Journal of the American Statistical Association* 65:797–807.

Milliken, G.A., and Johnson, D.E. 1984. *Analysis of Messy Data*,

*Vol. 1: Designed Experiments.* New York: Van Nostrand Reinhold Company.

Milliken, G.A., and Rasmuson, D. 1977. A Heuristic Technique for Testing for the Presence of Interaction in Nonreplicated Factorial Experiments. *The Australian Journal of Statistics* 19:32–38.

Scheffé, Henry. 1959. *The Analysis of Variance.* New York: John Wiley.

Schuurman, F.J., Krishnaiah, P.R., and Chattopadhyah, A.K. 1973. On the Distribution of the Ratios of the Extreme Roots to the Trace of the Wishart Matrix. *Journal of Multivariate Analysis* 3:445–453.

Tukey, John W. 1949. One Degree of Freedom for Non-additivity. *Biometrics* 5:232–242.

Ward, G.C., and Dick, I.D. 1952. Non-additivity in Randomized Block Designs and Balanced Incomplete Block Designs. *New Zealand Journal of Science and Technology* 33:430–435.

Yochmowitz, M.G., and Cornell, R.G. 1978. Stepwise Tests for Multiplicative Components of Interaction. *Technometrics* 20:79–84.

# Index